DSL For Dummies®

ILEC and CLEC Web Sites

ILEC/CLEC	Web Site
Ameritech (ILEC)	www.ameritech.com/products/data/adsl
Bell Atlantic (ILEC)	www.bell-atl.com/infospeed
Bell South (ILEC)	www.bellsouth.net/external/adsl
Covad Communications (CLEC)	www.covad.com
GTE (ILEC)	www.gte.com
NorthPoint Communications (CLEC)	www.northpointcom.com
Pacific Bell (ILEC)	www.pacbell.com/products/business/fastrak/adsl
Rythmns Net Connections (CLEC)	www.rythmns.com
Southwestern Bell (ILEC)	www.swbell.com/dsl
US WEST (ILEC)	www.interprise.com

DSL Flavors

DSL Flavor	What It Does
ADSL (Asymmetric DSL)	Delivers simultaneous high-speed data and POTS voice service over the same telephone line. Supports a range speeds from 1.5 Mbps to 8 Mbps downstream and 64 Kbps to 640 Kbps upstream.
UADSL (Universal ADSL)	Supports POTS. A variant of ADSL based on the G.Lite standard for 1.5-Mbps downstream and 384-Kbps upstream. UADSL is intended for the mass market including consumers, small businesses, and remote offices.
SDSL (Symmetric DSL)	Supports symmetric service at 160 Kbps to 2.3 Mbps. SDSL does not support POTS connections.
IDSL (ISDN DSL)	Offers an always-on alternative to dial-up ISDN service with a capacity of up to 144 Kbps.
RADSL (Rate-Adaptive DSL)	Offers automatic rate adaptation. Most ADSL is really RADSL, which allows actual data transmission rates to adjust to line conditions and distance. Downstream speeds can reach up to 8 Mbps, and upstream speeds can reach up to about 1 Mbps. RADSL supports both asymmetric and symmetric data transmission.
HDSL (High-data-rate DSL) HDSL-2	Supports symmetric service at 1.54 Mbps but does not support POTS. HDSL is the DSL service widely used for T1 lines. HDSL uses four wires (two pairs) instead of the standard two wires used for other DSL flavors. A new version, HSDL-2, provides the same speed capabilities as HDSL but uses only a single wire pair.
VDSL (Very-high-bit-rate DSL)	Supports up to 51 Mbps at very short distances. VDSL is the high-end member of the xDSL family.

...For Dummies®: Bestselling Book Series for Beginners

DSL For Dummies®

Cheat Sheet

Questions to Ask an ISP about DSL Internet Service

- What is the time commitment for your DSL service contracts?
- Do you offer price guarantees if DSL prices go down?
- Do you place usage restrictions on the amount of data going through the DSL connection?
- What are the terms of payment?
- What DSL CPE do you offer and support?
- Do you offer multiuser bridged and routed DSL service?
- If the ISP offers bridged service, do they have any special deals on security solutions such as proxy servers, Internet security appliances?
- How many routable IP addresses are included as part of the DSL service?
- How much do additional blocks of routable IP address cost per month?
- Do you place any restrictions on the operation of any TCP/IP application servers on the DSL connection?
- Do you place any restrictions on the number of computers using the DSL service?
- Do you offer any QoS (Quality of Service) guarantees?
- How much do ISP-hosted e-mail boxes cost per month?
- Do you support an e-mail server running on your LAN?
- Do you offer customer support 24 hours a day, 7 days a week?
- Do you charge for domain name registration services?

DSL Shopping Checklist

- Start your DSL availability search at the ILEC and CLEC Web sites.
- Shop around to compare DSL offerings from all the ISPs in your area.
- Look beyond just the price of the DSL to include all the variables of the Internet access package.
- Know what bandwidth you need before ordering your DSL service.
- Understand your DSL CPE options (and their prices) as part of your DSL service.
- Get a written quote from the ISP.
- Read the fine print in the Terms and Conditions contract.
- Ask about any special promotional deals.
- Calculate the total cost of your DSL service, including all start-up costs and monthly fees.
- Determine whether you plan to run any TCP/IP application servers on your DSL connection.

DSL CPE Vendor Web Sites

Vendor	Product Information
3Com	www.3com.com/solutions/dsl
Cayman Systems	www.cayman.com
Cisco	www.cisco.com
Efficient Networks	www.efficient.com
Flowpoint	ww.flowpoint.com
Netopia	www.netopia.com

...For Dummies®: Bestselling Book Series for Beginners

TM

...FOR DUMMIES

BESTSELLING BOOK SERIES

References for the Rest of Us! ®

Are you intimidated and confused by computers? Do you find that traditional manuals are overloaded with technical details you'll never use? Do your friends and family always call you to fix simple problems on their PCs? Then the *...For Dummies*® computer book series from IDG Books Worldwide is for you.

...For Dummies books are written for those frustrated computer users who know they aren't really dumb but find that PC hardware, software, and indeed the unique vocabulary of computing make them feel helpless. *...For Dummies* books use a lighthearted approach, a down-to-earth style, and even cartoons and humorous icons to dispel computer novices' fears and build their confidence. Lighthearted but not lightweight, these books are a perfect survival guide for anyone forced to use a computer.

> *"I like my copy so much I told friends; now they bought copies."*
>
> — **Irene C., Orwell, Ohio**

> *"Quick, concise, nontechnical, and humorous."*
>
> — **Jay A., Elburn, Illinois**

> *"Thanks, I needed this book. Now I can sleep at night."*
>
> — **Robin F., British Columbia, Canada**

Already, millions of satisfied readers agree. They have made *...For Dummies* books the #1 introductory level computer book series and have written asking for more. So, if you're looking for the most fun and easy way to learn about computers, look to *...For Dummies* books to give you a helping hand.

IDG BOOKS WORLDWIDE ®

DSL
FOR
DUMMIES®

DSL FOR DUMMIES®

by David Angell

Foreword by Reed Hundt

IDG Books Worldwide, Inc.
An International Data Group Company

Foster City, CA ◆ Chicago, IL ◆ Indianapolis, IN ◆ New York, NY

DSL For Dummies®

Published by
IDG Books Worldwide, Inc.
An International Data Group Company
919 E. Hillsdale Blvd.
Suite 400
Foster City, CA 94404
www.idgbooks.com (IDG Books Worldwide Web site)
www.dummies.com (Dummies Press Web site)

Library of Congress Catalog Card No.: 99-61342

ISBN: 0-7645-0475-4

Printed in the United States of America

10 9 8 7 6 5 4

1O/QS/QY/ZZ/IN

Distributed in the United States by IDG Books Worldwide, Inc.

Distributed by CDG Books Canada Inc. for Canada; by Transworld Publishers Limited in the United Kingdom; by IDG Norge Books for Norway; by IDG Sweden Books for Sweden; by IDG Books Australia Publishing Corporation Pty. Ltd. for Australia and New Zealand; by TransQuest Publishers Pte Ltd. for Singapore, Malaysia, Thailand, Indonesia, and Hong Kong; by Gotop Information Inc. for Taiwan; by ICG Muse, Inc. for Japan; by Norma Comunicaciones S.A. for Colombia; by Intersoft for South Africa; by Eyrolles for France; by International Thomson Publishing for Germany, Austria and Switzerland; by Distribuidora Cuspide for Argentina; by LR International for Brazil; by Galileo Libros for Chile; by Ediciones ZETA S.C.R. Ltda. for Peru; by WS Computer Publishing Corporation, Inc., for the Philippines; by Contemporanea de Ediciones for Venezuela; by Express Computer Distributors for the Caribbean and West Indies; by Micronesia Media Distributor, Inc. for Micronesia; by Grupo Editorial Norma S.A. for Guatemala; by Chips Computadoras S.A. de C.V. for Mexico; by Editorial Norma de Panama S.A. for Panama; by American Bookshops for Finland. Authorized Sales Agent: Anthony Rudkin Associates for the Middle East and North Africa.

For general information on IDG Books Worldwide's books in the U.S., please call our Consumer Customer Service department at 800-762-2974. For reseller information, including discounts and premium sales, please call our Reseller Customer Service department at 800-434-3422.

For information on where to purchase IDG Books Worldwide's books outside the U.S., please contact our International Sales department at 317-596-5530 or fax 317-596-5692.

For consumer information on foreign language translations, please contact our Customer Service department at 1-800-434-3422, fax 317-596-5692, or e-mail rights@idgbooks.com.

For information on licensing foreign or domestic rights, please phone +1-650-655-3109.

For sales inquiries and special prices for bulk quantities, please contact our Sales department at 650-655-3200 or write to the address above.

For information on using IDG Books Worldwide's books in the classroom or for ordering examination copies, please contact our Educational Sales department at 800-434-2086 or fax 317-596-5499.

For press review copies, author interviews, or other publicity information, please contact our Public Relations department at 650-655-3000 or fax 650-655-3299.

For authorization to photocopy items for corporate, personal, or educational use, please contact Copyright Clearance Center, 222 Rosewood Drive, Danvers, MA 01923, or fax 978-750-4470.

About the Author

David Angell is a principal in angell.com (`www.angell.com`), a technical communications firm in Boston. He has made a career of demystifying computers, telecommunications, and the Internet. David is author of 20 books, including *ISDN For Dummies* and *Microsoft Internet Information Server 4 For Dummies*. You can contact David at `david@angell.com`.

ABOUT IDG BOOKS WORLDWIDE

Welcome to the world of IDG Books Worldwide.

IDG Books Worldwide, Inc., is a subsidiary of International Data Group, the world's largest publisher of computer-related information and the leading global provider of information services on information technology. IDG was founded more than 30 years ago by Patrick J. McGovern and now employs more than 9,000 people worldwide. IDG publishes more than 290 computer publications in over 75 countries. More than 90 million people read one or more IDG publications each month.

Launched in 1990, IDG Books Worldwide is today the #1 publisher of best-selling computer books in the United States. We are proud to have received eight awards from the Computer Press Association in recognition of editorial excellence and three from Computer Currents' First Annual Readers' Choice Awards. Our best-selling ...*For Dummies®* series has more than 50 million copies in print with translations in 31 languages. IDG Books Worldwide, through a joint venture with IDG's Hi-Tech Beijing, became the first U.S. publisher to publish a computer book in the People's Republic of China. In record time, IDG Books Worldwide has become the first choice for millions of readers around the world who want to learn how to better manage their businesses.

Our mission is simple: Every one of our books is designed to bring extra value and skill-building instructions to the reader. Our books are written by experts who understand and care about our readers. The knowledge base of our editorial staff comes from years of experience in publishing, education, and journalism — experience we use to produce books to carry us into the new millennium. In short, we care about books, so we attract the best people. We devote special attention to details such as audience, interior design, use of icons, and illustrations. And because we use an efficient process of authoring, editing, and desktop publishing our books electronically, we can spend more time ensuring superior content and less time on the technicalities of making books.

You can count on our commitment to deliver high-quality books at competitive prices on topics you want to read about. At IDG Books Worldwide, we continue in the IDG tradition of delivering quality for more than 30 years. You'll find no better book on a subject than one from IDG Books Worldwide.

John Kilcullen
Chairman and CEO
IDG Books Worldwide, Inc.

Steven Berkowitz
President and Publisher
IDG Books Worldwide, Inc.

VIII WINNER
Eighth Annual Computer Press Awards ≥1992

IX WINNER
Ninth Annual Computer Press Awards ≥1993

X WINNER
Tenth Annual Computer Press Awards ≥1994

XI WINNER
Eleventh Annual Computer Press Awards ≥1995

IDG is the world's leading IT media, research and exposition company. Founded in 1964, IDG had 1997 revenues of $2.05 billion and has more than 9,000 employees worldwide. IDG offers the widest range of media options that reach IT buyers in 75 countries representing 95% of worldwide IT spending. IDG's diverse product and services portfolio spans six key areas including print publishing, online publishing, expositions and conferences, market research, education and training, and global marketing services. More than 90 million people read one or more of IDG's 290 magazines and newspapers, including IDG's leading global brands — Computerworld, PC World, Network World, Macworld and the Channel World family of publications. IDG Books Worldwide is one of the fastest-growing computer book publishers in the world, with more than 700 titles in 36 languages. The "...For Dummies®" series alone has more than 50 million copies in print. IDG offers online users the largest network of technology-specific Web sites around the world through IDG.net (http://www.idg.net), which comprises more than 225 targeted Web sites in 55 countries worldwide. International Data Corporation (IDC) is the world's largest provider of information technology data, analysis and consulting, with research centers in over 41 countries and more than 400 research analysts worldwide. IDG World Expo is a leading producer of more than 168 globally branded conferences and expositions in 35 countries including E3 (Electronic Entertainment Expo), Macworld Expo, ComNet, Windows World Expo, ICE (Internet Commerce Expo), Agenda, DEMO, and Spotlight. IDG's training subsidiary, ExecuTrain, is the world's largest computer training company, with more than 230 locations worldwide and 785 training courses. IDG Marketing Services helps industry-leading IT companies build international brand recognition by developing global integrated marketing programs via IDG's print, online and exposition products worldwide. Further information about the company can be found at www.idg.com. 1/24/99

Author's Acknowledgments

Although I'm the author of this book, it wouldn't have been possible without the help of many people along the way. At IDG Books, I want to thank Joyce Pepple, who offered me the opportunity to write this book. A special thanks to Susan Pink, my project editor, for another outstanding job. Big kudos goes to Danny Briere at TeleChoice for an outstanding technical edit.

Last but not least, my thanks to the following people and companies for providing essential support for this project. Brad Sachs at Flashcom; Michael Malaga, Reed Hundt, and Ann Zeichner; Sue Baelen, Elli Bradley, and Nina Croner at NorthPoint Communications; Louis Pelosi, James Naro, Mike Clark, and Mark Mazza at Covad Communications; Joe Peck, Lyn Gulbransen, Mike Frandsen, Devin Avery, Dave Casteel, and Ed Beuchot at Concentric; Erica Osmand at Stering Communications; Peter Borne, Mike Sampson, and Lori Hicks at Efficient Networks; Charles Sommerhouser at Sommerhouser Public Relations; Hidi Clark and Bill Muphy at Cayman Systems; Sharon Powers and Pamela Lee at Cisco Systems; Wendy Hawkin at FlowPoint; Rich Hatton and Kimberly Ellermeier at Netopia; Bruce Alter at 3Com; Alison Kenney at the Weber Group; Jeff Walhuter, Joan Ramussen, and Larry Plumb at Bell Atlantic; John Lentz at US West; Mark Mundt at Westell; Larry Woodard at Sonic Systems; Mary Kay Lies and Kevin Hong at MultiTech Systems; Jurgen Lison at Alcatel; Bill Rodey at ADSL Forum; Janice McCoy at the Massachusetts Public Utilities; Diana Helfrich at Copper Mountain; Brent Heslop at Bookware; and Jonathan Gaines at DMS Strategies.

Dedication

To my extended family: Paul and Kay Rennie; Paul, Holly, and Heather; David, Helen, and Karalyn; Karen, Scott, Drew, and Collin.

Publisher's Acknowledgments

We're proud of this book; please register your comments through our IDG books Worldwide Online Registration Form located at http://my2cents.dummies.com.

Some of the people who helped bring this book to market include the following:

Acquisitions, Editorial, and Media Development

Project Editor: Susan Pink

Acquisitions Editor: Joyce Pepple

Technical Editor: Danny Briere

Media Development Editor: Joell Smith

Associate Permissions Editor: Carmen Krikorian

Media Development Coordinator: Megan Roney

Editorial Manager: Mary C. Corder

Media Development Manager: Heather Heath Dismore

Editorial Assistant: Paul Kuzmic

Production

Project Coordinator: Karen York

Layout and Graphics: Linda M. Boyer, Angela F. Hunckler, Brent Savage, Renee L. Schmith, Kathie Schutte, Janet Seib, Kate Snell, Michael Sullivan

Proofreaders: Henry Lazarek, Nancy Price, Rebecca Senninger

Indexer: Betty Hornyak

Special Help
Bill Murphy, Suzanne Thomas

General and Administrative

IDG Books Worldwide, Inc.: John Kilcullen, CEO; Steven Berkowitz, President and Publisher

IDG Books Technology Publishing Group: Richard Swadley, Senior Vice President and Publisher; Walter Bruce III, Vice President and Associate Publisher; Steven Sayre, Associate Publisher; Joseph Wikert, Associate Publisher; Mary Bednarek, Branded Product Development Director; Mary Corder, Editorial Director

IDG Books Consumer Publishing Group: Roland Elgey, Senior Vice President and Publisher; Kathleen A. Welton, Vice President and Publisher; Kevin Thornton, Acquisitions Manager; Kristin A. Cocks, Editorial Director

IDG Books Internet Publishing Group: Brenda McLaughlin, Senior Vice President and Publisher; Diane Graves Steele, Vice President and Associate Publisher; Sofia Marchant, Online Marketing Manager

IDG Books Production for Dummies Press: Michael R. Britton, Vice President of Production; Debbie Stailey, Associate Director of Production; Cindy L. Phipps, Manager of Project Coordination, Production Proofreading, and Indexing; Shelley Lea, Supervisor of Graphics and Design; Debbie J. Gates, Production Systems Specialist; Robert Springer, Supervisor of Proofreading; Laura Carpenter, Production Control Manager; Tony Augsburger, Supervisor of Reprints and Bluelines

Dummies Packaging and Book Design: Patty Page, Manager, Promotions Marketing

◆

The publisher would like to give special thanks to Patrick J. McGovern, without whom this book would not have been possible.

◆

Contents at a Glance

Cartoons at a Glance

By Rich Tennant

"Troubleshooting's a little tricky here. The route table to our destination hosts include a Morse code key, several walkie-talkies, and a guy with nine messenger pigeons."

page 99

"THE PHONE COMPANY BLAMES THE MANUFACTURER, WHO SAYS IT'S THE SOFTWARE COMPANY'S FAULT, WHO BLAMES IT ON OUR MOON BEING IN VENUS WITH SCORPIO RISING."

page 257

"If it works, it works. I've just never seen network cabling connected with Chinese handcuffs before."

page 7

"I guess you could say this is the hub of our network."

page 185

"Just how accurately should my Web site reflect my place of business?"

page 281

Fax: 978-546-7747 • E-mail: the5wave@tiac.net

Table of Contents

Foreword

· ·

*H*ere we are, in the midst of a digital age: an age of promise, an age of possibility, and for many, an age of anxiety and apprehension. The embodiment of this era is, in all respects, the Internet. It is a technology every bit as revolutionary as the invention of the telegraph was more than 150 years ago.

In truth, the Internet is not a single technology, but a convergence of influences — personal computers, networking protocols, and even, I'm proud to say, the demise of monopolistic telecommunications policies. Nowhere is that more evident than in the expansion of Digital Subscriber Line technology.

As you'll soon read in this book, DSL is possible thanks to the efforts of both the familiar telephone companies and a new breed of carrier made possible by the Telecommunications Act of 1996. This competitive environment has helped make DSL's affordable, high-bandwidth access broadly available.

The Telecommunications Act helped change the terrain and prevented a replay of the problems that arose when ISDN was introduced: cumbersome, provider-driven processes that yielded little benefit for consumers. Protocols were regionally defined or came late, complicated pricing was dictated by a single source and based on a completely different type of communication: voice conversations.

Alternatively, the introduction of DSL has been highlighted by exciting market-driven developments. In just a few short years, dozens of companies serving DSL customers have been established, from equipment manufacturers to competitive local exchange carriers. Each of these companies has monitored and evaluated the marketplace, adjusting prices, availability, installation procedures, and more to respond to customer demands. This in itself is revolutionary.

DSL is also the ideal technology to test the Telecommunications Act. Based on copper, the dominant wiring medium, DSL is robust enough to warrant the creation of alternative carriers and cost-effective enough to be valued by both small businesses and residential consumers. After all, if the bandwidth and access are there, new services for home and business will not be far behind.

DSL For Dummies recognizes all the forces behind DSL's development, and excels not only in providing perspective but in helping us make the most of the technology. Beyond the historical, *DSL For Dummies* helps to clearly distinguish what's important to maximize the resources of this digital age.

Reed Hundt

—Former chairman of the Federal Communications Commission, and current member of the board of directors, NorthPoint Communications

Introduction

● ●

While you've been cursing the slowness of your analog modem or the high cost of your ISDN service, a new data communications service, called Digital Subscriber Line or simply DSL, has come into sight in the side-view mirror, and it's closer than it appears. Okay, I know your local telephone company has made all kinds of promises about high-speed Internet access, only to price it too high or deliver too little bandwidth, too late.

DSL is different. Why? Until the advent of DSL, the telephone companies had a monopoly on the delivery of data communications. On the low end of the bandwidth spectrum, telephone companies offered dial-up analog or ISDN; on the high end were the dedicated, leased line services of frame relay and T1 services. Millions of businesses, teleworkers, and consumers were left with a bad case of the bandwidth blues. Now, however, new competitive data communications companies are deploying DSL service and forcing the telephone companies to respond to the marketplace. The convergence of legislative-induced competition and the emergence of DSL technology are already resulting in affordable high-speed, always-on Internet access for the rest of us.

DSL breathes new life into those tired old telephone lines you use for voice and modems, transforming the same copper wire used for Plain Old Telephone Service (POTS) into a high-speed digital connection reaching speeds up to 8 Mbps. Speed isn't the only thing DSL brings to your Internet connection. DSL is an always-on data communications link, which means that you're always connected to the Internet 24 hours a day, 7 days a week — just like those guys with the expensive T1 lines. No more dial-up process or telephone company usage charges.

With a DSL connection, the double whammy of high-speed and always-on Internet connectivity translates into a cornucopia of new opportunities for working smarter and gaining competitive advantage. With a DSL connection, you can run your own Web server, get instant e-mail, do video conferencing, telecommute in the VPN (virtual private networking) fast lane, and a lot more.

About This Book

DSL For Dummies is your springboard for diving into the brave new world of DSL connectivity to the Internet. This hands-on guide presents your DSL options in a real-world context, cutting through the hype and technobabble to give you the specific information you need to be an educated consumer.

I explain the interrelated components that make up the entire ecosystem of DSL-based Internet access, including telecommunications, IP networking, and DSL hardware.

Foolish Assumptions

Assumptions can get you into a lot of trouble, but in writing this book, I made some assumptions about who you are and what you're probably looking for from DSL service. This book is for you if

- ✔ You've heard about DSL in some of its different incarnations (ADSL, SDSL, and so on), but you don't have a grasp of what DSL is all about and what it can do for you.

- ✔ You're using a PC or running a small LAN using Microsoft Windows 95, 98, or Windows NT (Server or Workstation).

- ✔ You've had it with dial-up modems that have made your Web experience the World Wide Wait.

- ✔ You've had it with high-priced ISDN from your local telephone company, and you're outgrowing its 128-Kbps limitation.

- ✔ You're a small business decision maker who has been frustrated at the lack of affordable, high-speed data connections to link your business to the Internet the way the big guys with their T1 lines do.

- ✔ You're on the lookout for new ways to gain a competitive advantage in today's business environment.

- ✔ You'd like to share a high-speed connection across several computers connected to a LAN. You probably have a LAN at the office and may even have, or plan to have, a LAN at home as the number of PCs in your home grows. You want an alternative to multiple telephone lines, each with its own modem and Internet account.

- ✔ You want to be an informed customer before you talk to an ISP about getting DSL service. You know that what you don't know can cost you.

- ✔ You don't have a lot of extra time. You want to know only the amount of technical mumbo-jumbo necessary to make the right decisions.

- ✔ Your business or organization has a limited budget, but you still want to enable people to collaborate remotely by using new real-time tools, such as video conferencing.

Conventions Used in This Book

This book's few conventions are shorthand ways to visually designate specific information from ordinary text. Here are the conventions:

- New terms are identified by using *italic*.

- Web site URLs (addresses) are designated by using a `monospace font`.

- Any command you enter at a command prompt is shown in bold and usually set on a separate line. Set-off text in italic represents a placeholder. For example, the text might read:

  At the DOS prompt, enter the command in the following format:

  ping *IPaddress*

  where *IPaddress* is the IP address of the remote host you want to check.

- Command arrows, which are typeset as ⇨, are used in a list of menus and options. For example, Start⇨<u>F</u>ind⇨<u>F</u>iles means to choose the Start menu, then choose the Find menu, and then choose the Files option. The appearance of an underline means you can press Alt plus the underlined letter instead of clicking the option with the mouse.

- Commands are shown with a plus sign, such as Ctrl+C. This means to hold down the Ctrl key while you press C.

How This Book Is Organized

DSL For Dummies has five parts that present material in a steady progression of what you need to know at the time you need to know it. Each part is self-contained but also interconnected to every other part.

Part I: Getting Comfortable with DSL

Part I lays the foundation for your DSL enlightenment. You discover how DSL service works and get help finding your way through the tangle of competing DSL technologies. You also find out about how TCP/IP and your choice of DSL hardware play a crucial role in determining what you can accomplish with your Internet connection. Finally, you unearth Internet problems associated with a DSL connection and are guided through a variety of affordable protection solutions.

Part II: Shopping for DSL

Part II gives you the game plan for navigating through the DSL shopping maze to get to the pipe dream. You discover TCP/IP application possibilities for taking advantage of high-speed, always-on DSL. You check out DSL ILEC and CLEC offerings and ISP partners. Wrapping up, you get the lowdown on what IP services, such as IP addresses, domain name services, Web hosting, and e-mail hosting, to consider as part of your Internet connection.

Part III: Connecting Your PC or LAN to DSL

Part III describes the mechanics of where the DSL connection meets your computer or LAN. If you want to connect multiple PCs to your DSL connection, you can find out how set up the network interface cards, cabling, and hubs or switches as the foundation for using a DSL bridge or router. As part of making your PC or LAN connection to the Internet through DSL, this part walks you through configuring Microsoft TCP/IP for Windows 95/98 and Windows NT computers. If you're using a DSL bridge or router to connect to the Internet, you can find out how to configure the TCP/IP settings for your network interface card. If you're using a DSL modem card or a USB modem, the section on working with Microsoft Windows Dial-Up Networking is for you.

Part IV: The Part of Tens

The Part of Tens provides information to help you further along your route to DSL enlightenment. Much of this information enhances and supports topics covered in previous chapters. You can find out my top ten reasons for getting DSL, ten ways to save money getting DSL service, and ten questions to ask an ISP.

Part V: Appendixes

Wait, there's more! As a bonus, *DSL For Dummies* includes two handy appendixes. Appendix A is an extensive listing of DSL products, services, and other resources for quick reference. Appendix B provides a useful glossary that defines the lingo and jargon of DSL.

What You're Not to Read

There is a lot of stuff you're not going to read in this book. You're not going to get into a technical jungle that goes off into splitting hairs. This book is about educating you on just what you need to get results — that is, the right DSL service and the right price. All technical discussions in this book play a supporting (not primary) role in helping you to make the right decisions.

Icons in This Book

DSL For Dummies includes icons that act as markers for special information. Here are the icons I use and what they mean.

 This icon signals nerdy techno-facts that you can easily skip without hurting your TCP/IP education. But if you're a technoid, you'll probably eat this stuff up. Enrich your mind or skip 'em if you like.

 This icon indicates nifty shortcuts or pieces of information that make your life easier. Here are the tips and tricks that will save you time, money, and aggravation.

 This icon lets you know that there's a loaded gun pointed directly at your foot. Watch out! Watch out! DSL may be the fast lane for Internet connections, but the road has potholes and hairpin curves. This icon points out these problems and tells you how to avoid or solve them.

Where to Go from Here

You are about to make your first move to the fast lane of Internet connectivity. Crack this book open and dive into Chapter 1 to get an executive summary of DSL and what it can do for you. From there, feel free to move about the book in a systematic linear manner or in random acts of information gathering. Either way, *DSL For Dummies* guides you through the world of high-speed, always-on DSL connections.

Part I

Getting Comfortable with DSL

In this part . . .

Today's modem-based Internet access is like being stuck in rush-hour traffic. You're stranded in the slow lane, and fast and affordable access to the Internet seems like a pipe dream. Digital Subscriber Line (DSL) promises to move you to the fast lane.

In Part I, you dive into the brave new world of DSL and what it can do for you. You get a solid grounding in how DSL service works and get help navigating through the maze of competing DSL technologies.

Next, you take the important step of acquiring some DSL data communication fundamentals. Then you move on to understanding how TCP/IP considerations and DSL hardware play a pivotal role in determining what you can and can't do with your Internet connection. Finally, I walk you through Internet problems associated with a DSL connection and guide you through a variety of affordable protection solutions.

Chapter 1

DSL: More than a Pipe Dream

. . . .

In This Chapter

▶ Searching for a cure for the bandwidth blues

▶ Discovering the DSL bandwidth sweet spot

▶ Exploring how a DSL connection changes everything about your Internet access

▶ Getting a reality check on DSL

. . . .

*F*or the millions of businesses and individuals stuck in the slow lane of analog modems or dial-up ISDN, fast and affordable access to the Internet has been a pipe dream. Welcome to the frustrated world of the bandwidth have-nots, where high-speed access has been denied for years by a monopolistic telecommunications industry.

Enter *Digital Subscriber Line (DSL),* a new data communications technology that promises to make high-speed bandwidth a reality for the rest of us. Why is DSL so promising? DSL, riding on the winds of change in the telecommunications industry, is a powerful data communications technology that works over regular telephone lines.

A Bad Case of the Bandwidth Blues

Bandwidth is the mantra of the Internet. Simply put, *bandwidth* is the capacity of any data communications link. The World Wide Web — today's GUI (graphical user interface) to the Internet — and the explosion of multimedia technologies have created a bandwidth-hungry environment that overwhelms today's modems. Anyone clunking around the Internet by using a dial-up modem over a system designed for voice more than a hundred years ago has felt the bandwidth blues.

While the Internet's evolution continues to speed along, the connectivity options for the majority of businesses, teleworkers, and consumers have languished in the hands of local telephone companies. As gatekeepers of data communications bandwidth, they have failed to deliver the kind of connections needed to support the relentless demands of today's Internet.

Until the advent of DSL, the bandwidth options offered by the telephone companies couldn't help but give you a bad case of the bandwidth blues. On the low end of the bandwidth spectrum are dial-up modems and ISDN; on the high end are the dedicated, leased line services of frame relay and T1 services.

The good news is that telephone companies are responding to the competition. Many of them are aggressively embracing DSL as a high-speed Internet service for the rest of us.

Modems, POTS, and the PSTN

For most Internet users, the analog modem is the current staple for data communications. A modem connection uses the same *POTS (Plain Old Telephone Service)* lines used for voice communications. This ubiquitous telephone system was designed to transmit the human voice as an analog waveform. Analog modems transmit digital computer information first by converting the digital data to analog signals. At the receiving end of the connection, a modem changes the analog signals back into the digital form used by computers.

The telephone lines from your home or office connect to a vast global telecommunications network referred to as the *Public Switched Telephone Network (PSTN)*. This network uses a complex system of computers to route calls based on telephone numbers. Although this system has been continually upgraded, its basic technology has been around for more than a hundred years. Analog-based data communications have inherent limitations that make today's 56 Kbps modems the end of the line for this technology.

The Internet strains the PSTN

Dial-up analog modem users and ISDN users have for years used the PSTN as the gateway to connecting to the more than 5,000 Internet service providers (ISPs) in the United States. As the Internet has exploded in popularity, the PSTN has been put under tremendous pressure from people using it for Internet access.

The problem is that the PSTN was designed for voice services, specifically for 3- to 7-minute voice calls (the general length of voice calls for the last hundred years). Internet sessions, however, average 30 minutes — much longer than the average voice call. These Internet calls tie up the PSTN and strain the system.

One of the most notable benefits of DSL is that it doesn't add more traffic to the PSTN for Internet access. DSL uses a parallel data communications network for high-speed data connections instead of routing data through the PSTN. As DSL service proliferates, it actually promises to shift data communications off the voice PSTN, which is something that the telephone companies will like.

ISDN: Too little, too late, too expensive?

ISDN (Integrated Services Digital Network) was the first attempt by the telephone companies to offer a higher speed, digital service to the mass market. ISDN delivers data at up to 128 Kbps over two 2 B (bearer) channels (at 64 Kbps each) using the same telephone wires used for POTS. The telephone companies have been deploying ISDN for several years, and it is widely available.

The telephone companies deploy ISDN as part of the PSTN, which makes most ISDN service a dial-up and metered service (although a few offer cost-effective unlimited rate ISDN services). Because ISDN is a premium service based on usage, it's expensive and impractical for Internet connectivity. Business ISDN service typically costs $100 to $350 for installation and anywhere between $50 to $330 (or higher) per month. Residential ISDN service is no bargain in most areas. For example, in the Boston area, Bell Atlantic charges about $50 a month for just the line and 1¢ per minute per B channel for a local data call. The usage rate might seem nominal, but it quickly adds up when you're using both B channels for several hours a day. Bills of a few hundred dollars per month are not unusual for residential ISDN service.

The record of the telephone companies in deploying ISDN is a real-world case study of how the telecommunications industry can muck up a new technology. Complex configuration and pricing as well as poor telephone company service also added to the mix in the marginalization of ISDN.

T1 and frame relay: High speed, high cost

The telephone company also offers leased lines, which are dedicated, high-speed links to the Internet installed between two points (called point-to-point service) to provide a dedicated (always-on) service. Users connecting to the Internet with a dedicated connection don't go through any dial-up process. Instead, connections are always ready and waiting for instant access. The leading types of leased and network services offered by the telephone companies are T1, fractional T1, and frame relay.

Although the cost of dedicated service has dropped, it remains prohibitively expensive for most individuals and small to medium-sized businesses. The local telephone companies continue to extract high prices for leased line services by using elaborate, outdated pricing models that have created a highly profitable business for them. Many companies buy alternative bandwidth from competitive access providers to get around such high prices, but prices are still expensive for most individuals and small to medium-sized businesses.

T1 lines are dedicated, leased lines that deliver data at 1.54 Mbps. The 1.54-Mbps pipeline is divided into 24 64-Kbps channels. These high-speed, digital links are created by using telephone lines with special equipment to handle higher data rates. The cost of a T1 line varies from $500 to $1100 for installation and $450 to $1000 (or higher) per month. Two charges are associated with T1 service for Internet access: one from the local telephone company, and the other from the Internet service provider.

A more powerful version of the T1 line is the T3 line, which is the equivalent of 28 T1 lines. Large companies and ISPs use this service for their backbone networks. A *backbone* net is a high-volume data communications link that carries data consolidated from smaller data communication links.

Ironically, the telephone companies have used DSL technology in the form of HDSL *(High-bit-rate Digital Subscriber Line)* for years to offer T1 line service. A new standards-based version of HDSL, called HDSL-2, promises to dramatically lower the cost of T1 service because it operates on only one pair of wires instead of the two required for HDSL. In addition, HDSL-2 minimizes interference when the wires run side-by-side with other data communication technologies.

Fractional T1 service is T1 service offered in fractional amounts based on 64-Kbps units, such as 256 Kbps, 384 Kbps, and 784 Kbps. A customer can start with fractional T1 service at one capacity level and upgrade as demand warrants. When a customer orders fractional T1 service, the carrier sets up a full T1 line but makes only the contracted bandwidth available. Fractional T1 service is expensive, although it is less expensive than full T1 service.

Frame relay — a public network offering from the telephone companies — is another popular form of leased line service. Frame relay is a data network service bundled with leased line access for transmitting data between remote networks. It's a network service built by local and long-distance telephone companies that acts like a private dedicated network. A typical 384-Kbps frame relay connection has installation costs from $700 to $1200 and a monthly cost from $550 to $850.

DSL: The Bandwidth Sweet Spot

Digital Subscriber Line (DSL) shares the same high-speed, always-on qualities of dedicated leased lines but at dramatically lower prices. DSL is actually a term used for a family of related telecommunications technologies that include such members as ADSL, SDSL, and IDSL. Each of these DSL flavors has unique bandwidth capabilities and limitations.

DSL service is so promising because it uses the same copper wiring as regular voice telephone service but bypasses the PSTN. DSL modems at each end of the line transform the telephone line into a high-speed digital connection.

The suite of DSL offerings promises to fill the chasm between slower dial-up modem and ISDN services and the fast (but expensive) T1 and frame relay services. DSL service is squarely targeted at this bandwidth sweet spot.

DSL brings a new level of Internet connections to businesses, teleworkers, and consumers. DSL brings the following three powerful data communication benefits to the rest of us:

- **Always-on connectivity.** There is no dial-up process, and the Internet is available for two-way data communications 24 hours a day, 7 days a week (commonly referred to as a 24 x 7 connection). These same attributes are inherent in traditional leased line offerings. An always-on connection translates to new opportunities for working smarter and gaining competitive advantage.

- **High speed**. DSL comes in a variety of speeds from 144 Kbps up to 52 Mbps. These bandwidth capabilities provide productive access to the full range of interactive content that the Internet has to offer. Most DSL service offerings also support bandwidth *scalability,* which means that you can increase your speed over time without incurring new start-up costs or buying new equipment.

- **Flat rate service.** The DSL link typically doesn't have usage-sensitive pricing, which means that the DSL connection can be used any time for as long as you need without incurring usage charges. The Internet access component of your DSL service, however, may have usage-based pricing after a certain threshold.

DSL is riding the winds of change

Digital Subscriber Line emanated from Bellcore, the research organization for the telephone companies formed after the breakup of AT&T. Telephone companies envisioned DSL as a way to deliver video-on-demand services to compete with the cable TV industry. These services never materialized in the United States, and DSL languished until bandwidth demand generated by the Internet explosion gave DSL a new lease on life.

In the hands of the telephone companies operating in their traditional monopolistic environment, DSL may have become another high-priced service outside the reach of all but the largest organizations. But change is swirling around the telecommunications industry, unleashing powerful competitive forces. DSL service deployment is at the forefront of taking advantage of this new competitive environment.

The Telecommunications Act and DSL

The new competitive environment driving the deployment of DSL was started by the passage of the Telecommunications Act in 1996. This legislation opens up the local telecommunications industry to competition, which ultimately gives consumers more choices at lower costs. Although by no means a perfect piece of legislation, it's already bearing fruit in the deployment of DSL.

Important parts of the Telecommunications Act are the provisions that require local telephone companies to allow competitors to resell their services and network elements on a nondiscriminatory basis. Local telephone companies, called *Incumbent Local Exchange Carriers (ILECs),* control the local telephone service infrastructure. The interconnection provisions of the Telecommunications Act require ILECs to open up their services and network elements to make them available to competitors. These competitors are called *Competitive Local Exchange Carriers (CLECs).* CLECs offering DSL data communication services exclusively are commonly referred to as *data CLECs* or *packet CLECs.*

CLECs make DSL service available to a given geographical area by installing equipment at the ILECs' facilities, called *central offices (COs).* COs are the terminus points on the telephone company side for the copper lines that reach out to every home and business in the United States. The competitive pressure on ILECs is forcing them to move more quickly and efficiently in deploying their own DSL services. Competition, however, isn't coming from only CLECs. The cable industry is also deploying high-speed Internet access services through their networks at competitive prices.

Meet the DSL family

Bellcore developed Digital Subscriber Line technology as a technique to filter out the background noise, or interference, on copper wires to allow clearer connections. This filtering process enables more data to move through regular telephone lines. DSL refers not to the telephone line itself but to the modems on each end of a line that convert the copper wiring into a high-speed digital line.

The generic term for the family of DSL services is *x*DSL, where the *x* is a placeholder for the letter used for a specific flavor of DSL. Each DSL flavor provides different data communications speed capabilities and limitations. Exact DSL service performance is based on the following factors:

  ✔ **The distance between a user's premises and the telephone company's central office**

  ✔ **The DSL equipment used at both ends of the connection**

  ✔ **The service offerings from ILECs and CLECs**

The distance of your premises from the CO is one of the most significant factors in determining availability of DSL service. Many DSL flavors have distance restrictions of up to 18,000 feet from the CO. Some DSL flavors, however, go beyond these distance limitations, all the way up to 30,000 feet.

Table 1-1 describes the key attributes of the DSL family of technologies. The leading DSL services being deployed and targeted at the bandwidth sweet spot are ADSL, G.Lite, SDSL, RADSL, and IDSL. Chapter 3 goes into more specifics about each form of DSL.

Table 1-1   The *x*DSL Family of High Bandwidth Services

DSL Flavor	What It Does
ADSL (Asymmetric DSL)	Delivers simultaneous high-speed data and POTS voice service over the same telephone line. Supports a range of speeds from 1.5 Mbps to 8 Mbps downstream and 64 Kbps to 640 Kbps upstream.
G.Lite	Supports POTS. A variant of ADSL based on the G.Lite standard for 1.5 Mbps downstream and 384 Kbps upstream. G.Lite is intended for the mass market, including consumers, small businesses, and remote offices.
SDSL (Symmetric DSL)	Supports symmetric service at 160 Kbps to 2.3 Mbps. SDSL does not support POTS connections.
IDSL (ISDN DSL)	Offers an always-on alternative to dial-up ISDN service with a capacity up to 144 Kbps.
RADSL (Rate-Adaptive DSL)	Offers automatic rate adaptation. Most ADSL is really RADSL, which allows actual data transmission rates to adjust to line conditions and distance. Downstream speeds can reach up to 8 Mbps, and upstream speeds reach up to around 1 Mbps. RADSL supports both asymmetric and symmetric data transmission.
HDSL (High-bit-rate DSL) HDSL-2	Supports symmetric service at 1.54 Mbps but does not support POTS. HDSL is the DSL service already widely used for T1 lines. HDSL uses four wires (two pairs) instead of the standard two wires used for other DSL flavors. A new version, HSDL-2, provides the same speed capabilities as HDSL but uses only a single wire pair.
VDSL (Very high-bit-rate DSL)	Supports up to 51 Mbps at very short distances. VDSL is the high-end member of the *x*DSL family.

When people talk about data rates and DSL, they use the terms upstream and downstream. The *downstream data rate* refers to the speed data travels from the Internet (or other remote network) to your local computer or network. The *upstream data rate* refers to the speed data travels from your local computer or network to the Internet or other remote network.

Asymmetric DSL means that the downstream speed is higher than the upstream speed. For example, an ADSL connection might have a downstream speed of 1.5 Mbps and an upstream speed maximum of 384 Kbps. *Symmetric DSL* means that the downstream and upstream speeds are the same. DSL service rates are typically represented as downstream/upstream. For example, an asymmetric offering might be noted as 1.5 Mbps/384 Kbps, and a symmetric offering might be noted as 384 Kbps/384 Kbps.

The great DSL bypass

Communications over the Public Switched Telephone Network (PSTN) are handled by circuit switches, which route telephone traffic from one destination to another. In a *circuit-switched system,* a temporary circuit, or path, is established between two points by dialing. The circuit is terminated when one end of the connection hangs up, which sends a disconnect signal. Dial-up analog modem and ISDN users connect to the PSTN right along with voice communications.

The expensive and time-consuming process of adding new services to the vast PSTN infrastructure was one of the main reasons ISDN service remained expensive and took years to deploy. DSL service lines bypass the circuit-switching infrastructure of the PSTN and terminate at the CO, where they are routed to their own data networking equipment called DSL Access Multiplexers (DSLAMs). From the DSLAM, the data moves to a backbone network that connects to your Internet service provider or corporate network.

DSL can ride over the same telephone line as POTS, and many ILECs are deploying DSL in that fashion. For example, using ADSL or G.Lite, the high-speed data goes to the DSLAM, but the POTS service is split off and sent to the PSTN switch, as shown in Figure 1-1.

The CLECs are mostly deploying data-only circuits, which do bypass the PSTN switch, as shown in Figure 1-2. Because DSL bypasses the PSTN switching infrastructure, it can be deployed much more easily and rapidly than any data communications offering routed through the PSTN.

Availability and price

At this early stage of the DSL life cycle, service offerings and prices are all over the map — both literally and figuratively. These are turbulent times for DSL service deployment and prices.

DSL service is being rolled out by ILECs, CLECs, and ISPs acting as CLECs. Because DSL is deployed on a CO-by-CO basis, we have a patchwork of DSL service availability. DSL deployment in the first round is focused on large metropolitan markets because these markets represent the largest concentration of potential DSL customers.

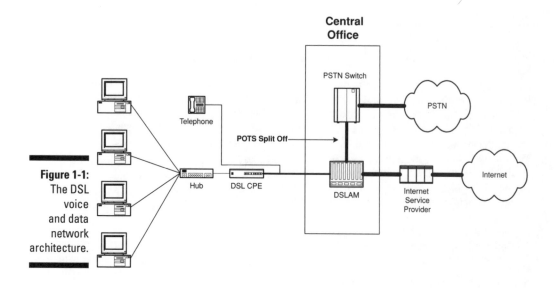

Figure 1-1:
The DSL
voice
and data
network
architecture.

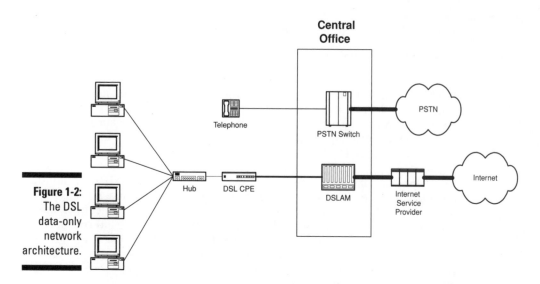

Figure 1-2:
The DSL
data-only
network
architecture.

Most ILEC DSL Internet service offerings are based on ADSL (Asymmetric DSL). ILECs are using ADSL because it allows customers to share the same line for both high-speed data communications and POTS. Most data CLECs, such as NorthPoint Communications and Covad Communications, are deploying SDSL (Symmetric DSL) and IDSL (ISDN DSL). SDSL is grounded in the same technology used for T1 lines, so it can be widely deployed without creating the kinds of technical problems caused by ADSL service.

A real-world DSL deployment

While I was writing this book, the leading data CLECs, NorthPoint Communications and Covad Communications, had already deployed DSL service in the Boston metropolitan area. Because these CLECs sell their DSL service wholesale to ISPs, shopping for DSL service for most users in the Boston area involves contacting ISPs that are partnering with CLECs. Interestingly, prices for the same 384-Kbps DSL service initially ranged from $199 a month to more than $700 a month, depending on the ISP. Setup and equipment charges ranged from $425 to $1500 and higher.

Leading the charge in the more affordable DSL offerings were new ISPs moving into the Boston area and riding exclusively on the CLECs' DSL deployment. These ISPs, such as Flashcom and Concentric, moving from the huge, competitive California market, consistently offered substantially lower prices than incumbent ISPs serving the Boston area. From the consumer standpoint, this competition was good news.

As the low-price leaders established themselves in the Boston market, most ISPs dramatically cut their DSL service prices. As a result, the spread between low and high prices narrowed considerably.

The ILEC serving the Boston area, Bell Atlantic, wasn't planning any DSL deployment until mid-1999. Bell Atlantic offerings include speeds of 640 Kbps/90 Kbps, 1.6 Mbps/90 Kbps, and 7.1 Mbps/680 Kbps.

Pricing for DSL service depends on a variety of speed and configuration options. The foremost factor in pricing is speed: More speed costs more. The more bandwidth you order, however, the better the value because one-time setup charges typically remain fixed regardless of the speed you choose within a specific DSL flavor.

Prices for DSL service range from around $40 a month up to $750 per month. Most DSL service offerings targeted at teleworkers and consumers range from $39 to $89 per month. DSL offerings for businesses are typically in the range of $100 to $500 a month. These prices typically include Internet access as well as the DSL connection. Perhaps the biggest expense in getting DSL service is the start-up cost, which typically includes a one-time charge for the DSL service, the Internet access account, and a DSL modem. These one-time charges can range anywhere from $300 to $700.

CLECs are aggressively rolling out DSL services, often beating ILECs in many markets. In most major metropolitan markets, CLECs are offering DSL services through ISPs. ILECs are a mixed bag in deploying DSL services. Pacific Bell (owned by SBC Communications), US WEST, and GTE are leading the ILEC pack in deploying DSL service. Ameritech is the slowest in DSL deployments, with Bell Atlantic, Bell South, and Southwestern Bell (owned by SBC Communications) falling somewhere in between.

Chapter 9 provides an extensive listing of DSL service providers.

A DSL Connection Changes Everything

Like expensive dedicated services, DSL is an always-on connection, which means that you're open for business 24 hours a day, 7 days a week. The always-on, high-speed capabilities of DSL enable you to do many things, including the following:

- **Cruise the Web with the top down.** At a very basic level, Internet connections through DSL become a thrill. The always-on DSL connection means that double-clicking the Microsoft Internet Explorer or Netscape Communicator icon instantly delivers the Internet to your desktop. No more squealing modem connections that seem to take forever. No waiting for Web page downloads. In fact, you're connection becomes so fast, you begin to see how slow different Web servers are in delivering information to you. You may find yourself waiting for a Web server to deliver data instead of waiting for your connection to do so — there's a twist.

- **Harness Internet voice and video communications**. In the world of low bandwidth and dial-up modems, using IP voice and video conferencing simply wasn't a viable communications solution. The high-speed capabilities of DSL make IP voice and video conferencing a practical, everyday tool. These real-time communications technologies become real-world solutions that can save you money and time. The always-on part of the DSL connection means that you're accessible to anyone connected to the Internet. An IP (Internet Protocol) address becomes the equivalent of a telephone number that is always available for incoming or outgoing calls. An *IP address* (also referred to as an *Internet address* or a *host address*) is a unique 32-bit number, such as 199.232.255.113, that identifies a specific computer or other network device on a TCP/IP network.

- **Support virtual organizations.** DSL enables every participant in a virtual enterprise to be both a user and a provider of information. You can link up geographically dispersed groups to operate as a virtual team. Electronic whiteboards combined with video conferencing over a DSL connection can create a powerful set of real-time, interactive tools that bring dispersed teams of people to new levels of communication.

- **Push and pull your applications.** The Internet is rapidly becoming the software distribution medium of choice. Entire programs can be quickly downloaded from a software publisher's Web site. Using push and pull technologies, software vendors can send updates to your software automatically. Push technologies can deliver customized news directly to any desktop on your LAN (local area network). This means that news, stock quotes, and other information will be constantly streaming down to your business.

✔ **Move content rapidly.** The size of programs and files grows almost exponentially as computer applications become more sophisticated. Desktop publishers, multimedia developers, software publishers, and a host of others can deliver digital content economically through DSL connections either as a download or by providing file servers. Downloading a 72MB file, for example, takes 25 minutes at 56 Kbps, 10 minutes at 128 Kbps, and 48 seconds at 1.5 Mbps.

✔ **Telework in the fast lane.** *Teleworkers* include telecommuters or anyone who uses information and telecommunications technologies to replace travel. DSL is a powerful tool for making you feel like you're sitting at a desk in the corporate office even though you're working remotely. Using virtual private networking (VPN) over the Internet, telecommuters can do work on the Internet as well as link up to computers on a company

Teleworkers of the world unite!

Teleworkers are those who use data communications as an alternative to travel, such as telecommuters, home-office workers, and after-hours corporate employees.

About 60 percent of the United States workforce is in the information workforce category. The *information workforce* consists of those who make substantial portions of their income by creating, manipulating, transforming, or transmitting information or operating information machines (computers). An estimated 80 percent of the information workforce could telecommute at least some of the time — that's about 48 percent of the total workforce. The use of data communications technologies, such as DSL, enables information workers to re-create the support services of the traditional office.

The gridlock of daily commuting, according to AAA (American Automobile Association) costs companies and employees 2 billion hours annually, as longer and longer commutes become reality. Telecommuters report a minimum of a 15 to 20 percent increase in productivity when they work outside the office, a figure that their managers corroborate. Many studies have confirmed productivity increases in teleworkers ranging from 15 to 50 percent. A Booz, Allen &

Hamilton study showed that about 25 percent of managers' time in the office is unproductive; for employees, the percentage is even higher. It should come as no surprise, then, that more than 80 percent of all organizations will have at least 50 percent of their staff engaged in telecommuting by the end of 1999, according to the Gartner Group.

A study by Arthur D. Little Associates found that if 10 to 20 percent of automobile commuters switched to telecommuting, savings would total $23 billion annually. These savings would be realized by eliminating 1.8 million tons of regulated pollutants, saving 3.5 billion gallons of gas, freeing 3.1 billion hours of personal time from reduced congestion and automobile trips, and reducing maintenance costs for the existing transportation infrastructure by $500 million.

The number of teleworkers is growing even faster than the number of telecommuters. More than 11 million teleworkers are in the United States today, a 30 percent increase from 1995 levels. In Silicon Valley alone, more than 900,000 of the area's 3 million workers already perform at least some work at home, and this number is rising 10 to 15 percent annually.

network. DSL-based remote access enables sophisticated and bandwidth-hungry applications, such as video conferencing and Web-based collaborative computing, to become everyday teleworker tools.

✔ **Level the data communications playing field for small to medium-sized businesses.** Small businesses are already heavy users of the Internet, but because of the high cost of traditional dedicated connections (T1 and frame relay), they could not become players in providing services for Internet users. A DSL connection enables smaller businesses to do the same types of things with their Internet connection that large companies do, such as running Web and e-mail servers.

DSL Reality Check

DSL promises big changes for the way the rest of us connect to the Internet or corporate networks, but DSL is not a yellow brick road. Deploying DSL is a huge undertaking affected by a variety of technical, financial, and political factors. No honest coverage of DSL would be complete without alerting you to the elements that make up the dark side of DSL technology and its deployment.

This section presents an overview of the main trouble spots in DSL today. Keep in mind, however, that DSL is a fast-moving technology, and many of these problems will be resolved over time. Detailed specifics on these problems and how they can affect you are covered in relevant chapters throughout this book.

Subject to change without notice

DSL is a new data communications service that is in a state of constant change. DSL service packages, user-equipment options, and service pricing are all subject to change without notice. Almost everything goes in terms of today's DSL service offerings. Making the right decision about your DSL service takes patience and persistence. Most of the trends are good for consumers in terms of lower service and hardware prices as well as better service. However, DSL — like any new technology — will have growing pains.

Keeping your distance

DSL is distance sensitive. Most DSL flavors have a distance limitation of 12,000 to 18,000 feet between the customer's premises and the central office, but these distance limitations are being extended with new technology improvements. This distance limitation has to do with the higher frequencies used by DSL to enable higher data speeds. DSL service has inherent limitations based on the distance between a user's premises and the central office servicing a given area. This distance is not as the crow flies, but the snaking path that the copper link takes. By most estimates, 60 to 80 percent of the United States population lives close enough to a central office to take advantage of the more-popular DSL technologies.

Please stand by, we're experiencing technical problems

In addition to distance limitations, a variety of technical issues affect DSL deployment. For example, certain flavors of DSL, such as ADSL, are subject to *crosstalk*, or interference, when bundled in the same cable with other DSL flavors or older data communications technologies used for T1 lines.

Fragile competition

Even with the legislative push of the Telecommunications Act, ILECs continue to wield tremendous power in shaping DSL service and deployment in their respective service areas. They have the legal and lobbying resources to challenge the government and the implementation of all the provisions of the Telecommunications Act. The ILECs have a great deal to fear from the deployment of DSL because it threatens to cannibalize their profitable leased line and network services (T1 lines and frame relay).

ILECs don't make it easy for CLECs for deploy DSL service. The process for CLECs to get access to an ILEC's CO space and local loops is a long and arduous journey. From the DSL consumer perspective, the success of CLEC competition is critical to ensuring competitive prices and service options. Compared to ILECs, data CLECs are generally smaller, faster moving, technology-driven companies that use their speed and capabilities to deploy innovative data communications solutions.

A key element of the Telecommunications Act, however, is keeping most ILECs on the path of opening up their networks. ILECs may enter the lucrative long-distance telephone service market if they can convince the FCC (Federal Communications Commission) that they are complying with the interconnection provisions.

The powerful ILECs, the FCC, state regulatory agencies, and CLECs are involved in constant wrangling that could change the effect of the Telecommunications Act. Competition is also going on between ILECs and the government as well as between ILECs and CLECs. At this early stage of competition in local telecommunications, the provisions of the Telecommunications Act are fragile.

Waiting for interoperability

A *protocol* is a set of rules for data communications. A *standard* is a set of detailed technical guidelines used to establish uniformity. Protocols and standards create an environment of universal capability. A pivotal milestone in the deployment of affordable, off-the-shelf DSL equipment (referred to as *CPE* for Customer Premises Equipment) is the emergence of standards. DSL equipment interoperability is an essential factor in lowering the cost of CPE and increasing consumer choice. With multiple vendors using the same standards for DSL service, consumers can choose the best CPE for their needs.

A new international standard for ADSL, called G.Lite, has emerged from the International Telecommunications Union (ITU). The G.Lite standard can deliver 1.5 Mbps downstream and 384 Kbps upstream. This standard promises to enable multiple vendors to offer — through the computer retail channel — DSL modems that will work with different DSL providers.

Other forms of DSL lack standards, which means that your DSL equipment options for many flavors of DSL are limited. What DSL equipment you can use at your premises is restricted to the equipment supported by the ILEC or CLEC providing the DSL service. As a stopgap measure, many DSL modem vendors have interoperability agreements with vendors providing the DSLAM equipment at the CO.

The lack of DSL standards doesn't preclude your use of DSL now because many DSL modems can be upgraded by using firmware upgrades downloaded from a vendor's Web site. This means that the DSL equipment you buy today has a good chance of not becoming obsolete with the emergence of standards, unless you change your DSL service. The risk remains, however, that something in your DSL connection might change and make your equipment obsolete. In addition, the lack of standards means that the equipment you're buying now will be more expensive than standards-based CPE sold in a competitive market.

Bandwidth shell game

The ISP business is about leveraging bandwidth to get the most bang for the buck. The urge to oversubscribe is a compelling economic issue with ISPs, and they all play the bandwidth game. This oversubscribing may become intense with DSL deployment because an ISP offering relatively inexpensive high-speed connections to customers won't be able to afford the high-cost backbone network to adequately support the DSL links to and from the Internet. An ISP estimates the amount of data traffic it will experience and furnishes a backbone based on that estimate.

The bandwidth you're buying to access the Internet through DSL is really the speed supported by the DSL line from your premises through the DSLAM at the CO to the DSL provider's (CLEC or ILEC) backbone. The connection between the ISP and the DSL provider's backbone and the ISP to the Internet may be too small to support the volume of traffic. If the capacity of this back end is low compared to the capacities the ISP has sold to its customers, a bottleneck occurs.

What's on Cable?

No discussion of DSL service is complete without covering its main competitor, cable modem service. The second largest network in the United States is cable television, which passes 90 percent of all homes. The cable companies sit on a huge untapped source of bandwidth for Internet connections. Unfortunately, only about 15 percent of cable systems can support high-speed, two-way connections to the Internet. To handle the two-way traffic used in Internet connections, cable companies must invest in expensive new infrastructure to upgrade their systems.

Cable modem service is primarily a residential Internet access service because most residential areas are wired for cable service, but most business locations are not. Approximately 22 million homes in the United States have access to high-speed cable modem service (called *homes passed*), and as of the end of 1998, cable modem subscribers numbered between 500,000 and 700,000.

Cable modem service can be a great bandwidth value, but it can also suffer from its own success. Cable service is a shared connectivity medium much like a local area network. The cable system architecture connects hundreds or thousands of homes into large Ethernet networks. As more users are connected to the cable network in a given area, the throughput of the connection slows down for everyone.

These shared networks also create serious security risks for any PC connected to the cable network. In Microsoft Windows, if you share a resource, such as a drive, it might be available to other Windows users connected to your neighborhood cable network. Some cable companies, however, have instituted solutions to this problem.

Although cable modem service is a fast, always-on connection, you can't take full advantage of these capabilities. Most cable modem service providers don't provide static IP addresses or support DNS services, which means that you won't be able to run any type of server, such as a Web or e-mail server, that uses a domain name address. In addition, most cable modems are designed to connect to only a single computer through a network interface card *(NIC)*. Note, however, that you can get around this limitation by using a proxy server or an Ethernet-to-Ethernet router to share the cable connection across a LAN.

Cable modem service might take the consumer market

A recent Forrester Research report, "Broadband Hits Home," predicts that by 2002, a quarter of the households in the United States with Internet access will use broadband connections. Of those 16 million homes, 80 percent will be connected through a cable company, and the remaining 20 percent will use DSL technology provided by the ILECs. Because data CLECs aren't targeting the DSL consumer market, they're not factored into this report.

Forrester expects cable data services to grow from the 350,000 current subscribers to more than 2 million by the end of 1999, driven by lower prices for cable modems and increased consumer awareness. The phone companies will be held back by competing technologies, lack of standards, and high equipment costs. Consumers will be disappointed by the telco offerings because of their initial price/performance ratio and service limitations. For my two-cents worth, I wouldn't write off DSL and the telephone companies in the broadland race to the home — it's too early.

Should you check out cable modem service?

If your Internet access and bandwidth needs fit the constraints of cable modem service, you should consider cable service (if it's available in your area). Cable modem service may be a better deal than comparable DSL consumer offerings from your local ILEC. For example, MediaOne offers MediaExpress cable modem service in the Boston area for $30 to $50 a month, which includes the connection and Internet service. For this flat monthly price, you can connect a single PC to the Internet by using an always-on connection. The installation fee is $100 or is $149 if you need a network adapter card, which connects your PC to the cable modem. The MediaOne cable modem delivers a downstream speed of 1.54 Mbps and an upstream speed of 300 Kbps. Bell Atlantic's InfoSpeed service costs $59.95 a month for 640 Kbps/92 Kbps.

Cable competition can't hurt

Cable operators offering cable modem service represent a major competitive force to the consumer DSL offerings of the telephone companies. Cable companies, with their aggressive pricing and early leadership in taking a new approach to installing a high-bandwidth service, have forced the telephone companies to do things differently.

Cable companies were the first to buck conventional wisdom in terms of going to the customer's site and getting the entire Internet connection up and running. They eliminated the apartheid between computing and data communications, which has been the way telephone companies have treated their data communications service.

The cable industry has had a year's head start in deploying affordable, high-speed Internet access. Cable companies have delivered their cable modem service in a user-friendly way that challenges the telecommunications industry to do the same. When a user orders cable modem service, they come to your home, connect the cable modem to your PC through an Ethernet connection, and even install a network interface card if you don't have one. They do all the cabling. When they leave, you're up and running. The bottom line is that cable modem service is here to stay, and if it fits your needs, go for it.

Chapter 2

The Telecom Side of DSL

. . . .

In This Chapter

▶ Understanding DSL's telecommunications habitat

▶ Mastering the lingo of switches, central offices, and local loops

▶ Entering the world of the DSL network

▶ Checking out the equipment at the end of the DSL line

▶ Evaluating DSL from CLECs

▶ Grading DSL from ILECs

. . . .

DSL is the data communications conduit between you and your Internet service provider or corporate network. Although the DSL part of your connection is transparent, its operation plays a big role in defining your DSL service options. This chapter takes you on a guided tour of the telecommunications side of your DSL connection.

Welcome to the Telecommunications Jungle

The telecommunications industry is in the throes of a massive upheaval. A convergence of legislative, technological, and competitive forces is fueling changes in the way telecommunications services are delivered. With change comes large-scale confusion for telecommunications providers, Internet service providers, and consumers. In the long run, however, the unleashing of competitive forces into the monopolistic telecommunications industry is in the consumer's best interest.

DSL service deployment is heavily influenced by these changes. DSL isn't only a new data communication technology; it's also spearheading the new era of competition in the telecommunications industry. Understanding how the telecommunications industry operates is an essential starting point in your journey toward DSL enlightenment.

Gatekeepers of the last mile

AT&T, which was known as the Bell System, was once the only telephone company for most of the United States. In 1984, the United States government forced AT&T to divest itself of its local telephone companies. The local operations of AT&T were broken up into Regional Bell Operating Companies (RBOCs). One of the most significant changes mandated by the breakup of AT&T was the redefinition of the nation's telecommunications system into local exchange carriers (LECs) and interexchange carriers (IXCs).

An LEC is the telephone company that provides local telephone service. An IXC is a long-distance telephone company. This forms the basis of the telephone system today. Every home and business in the United States with telephone service goes through an LEC. Long-distance calls are routed from the LEC to the long-distance telephone company and then to the LEC at the destination.

LECs include both RBOCs (Regional Bell Operating Companies) and independent telephone companies (ITCs). There are close to 1400 independent telephone companies, including some big players such as GTE, Frontier Telecommunications, and Southern New England Telecommunications (SNET). Most ITCs, however, are local telephone companies servicing rural areas. Although these ITCs supply telephone service to only 15 percent of the telephone customers in the United States, their service area encompasses half its geographical area.

RBOCs are the dominant players in local telephone service. You know these RBOCs as Ameritech, Bell Atlantic, BellSouth, Pacific Bell, Southwestern Bell, and US WEST. LECs provide the connections to almost every telephone user in the United States. They control the copper wiring that connects almost every home and business within a given geographical area to an LEC facility called a central office (CO). This wiring, called a *local loop,* consists of the twisted-pair copper wiring used for POTS (Plain Old Telephone Service) connections between central offices (COs) and customer sites. DSL service comes into play where these local loops connect to central offices. The control of the last mile by the LECs operating as monopolies has been the largest single factor in the high cost of bandwidth for data communications.

Changing the rules

The ultimate goal of the Telecommunications Act of 1996 is to open up local telecommunications services to competition. In the parlance of the Telecommunications Act, LECs became known as *Incumbent Local Exchange Carriers (ILECs)* because they control the local telephone service infrastructure (COs and local loops). The new telecommunications players are called *Competitive Local Exchange Carriers (CLECs).*

The carrot and the stick

As a trade-off for ILECs giving up some of their monopoly in the local telephone service market, they were offered the carrot of gaining access to the lucrative long-distance market. Currently, ILECs are not allowed to sell long-distance service within their regions until the FCC is satisfied that their networks, on a state-by-state basis, have opened up their services to CLECs. However, ILECs will be allowed to enter the long-distance market three years after the passage of the Telecommunications Act of 1996, if an agreement hasn't been signed by then.

At the heart of DSL deployment by CLECs are the Telecommunications Act's provisions covering unbundled access and colocation. These network elements include central office space, transmission between COs in a metropolitan area, and wiring systems used to deliver telecommunications services to customers. Using the interconnection provisions of the Telecommunications Act, CLECs can install their DSL equipment (called Digital Subscriber Line Access Multiplexers, or DSLAMs) at the COs owned by ILECs. CLECs then lease local loops that terminate at the CO to provide DSL service to customers.

Through the regulatory maze

The Federal Communications Commission (FCC) along with 50 state regulators define the rules for the delivery of telecommunications services. Telecommunications oversight, policy, and regulation at the federal level is the province of the FCC. These functions at the state level are performed by each state's regulatory agencies, typically called public utility commissions (PUCs). These state agencies add another layer of regulations for telecommunications at the state government level and exercise regulatory powers over ILECs and CLECs.

The Telecommunications Act has created ongoing confusion on jurisdiction between the FCC and state public utility commissions on the control of in-state pricing of ILECs and CLECs. The details of the implementation of the Telecommunications Act and the enforcement of its provisions were left, for the most part, to the FCC. The FCC has made a number of rulings related to the implementation of the Telecommunications Act, and challenges to many of them are tied up in the courts. The main message here is that the Telecommunications Act is an evolving framework subject to changes, as challenges in courts become formalized.

In telephone industry lingo, prices are called tariffs. More specifically, a *tariff* is a published price that sets the allowed rate for telecommunications services and equipment. Tariffs are filed in each of the 50 states by way of public utility commissions (PUCs) or similar agencies. These governmental agencies are chartered to look after the public interest within the recognized monopoly power of essential utilities. As a check on state regulatory agencies, the Telecommunications Act disallows any state agency from prohibiting any qualified entity from providing interstate or intrastate telecommunications service.

Inside the Telecommunications Network

DSL service is a data communications service that operates in the realm of the telecommunications network of central offices and local loops but uses them differently than the way they're used for voice communications. The COs and local loops become the gateway to a data communications network as well as the voice communications network. The PSTN (Public Switched Telephone Network) is a wide area network on a global scale. Just about every business and home connects to this network through twisted-pair copper wires, which carry both voice and data communications.

All loops lead to the central office

The central office (CO) is the front line of the telecommunications network, as shown in Figure 2-1. Think of a CO as a network node on the PSTN. All local loops from every home and business terminate at a CO that services a specific geographical area. The term *local loop* is a generic term for the connection between a customer's premises and the CO.

The United States has more than 19,000 COs, which terminate more than 200 million local loops. (An estimated 700 million local loops exist worldwide.) Large metropolitan areas typically have a number of COs, which can be in an entire building or a room in a city building. These central offices connect to other COs for local calling or to other switching facilities for long-distance calls.

The common symbol used to represent any wide area network is a cloud, as you can see in Figure 2-1. The concept of a cloud relates to the fact that the operation of the network is hidden from the user.

Keeping tethered through local loops

The local loop connects your premises to the PSTN through the CO. Local loops are also called *access lines* because they enable users to access the CO. The majority of local loops are within 18,000 feet of a CO.

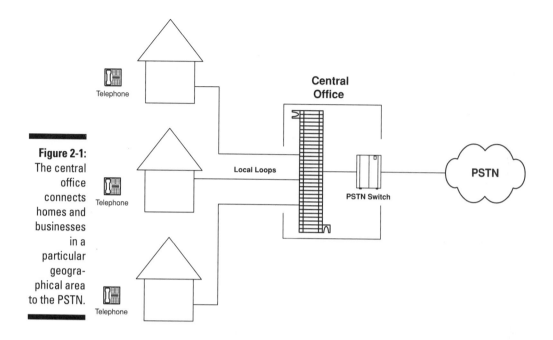

At the CO, local loops end up in the Main Distribution Frame (MDF), which is the wiring center for local loops. For POTS services, the wiring from the MDF is connected to the DSLAM and PSTN switch. From the PSTN switch, wiring is carried on telephone poles or buried in the ground until it reaches a local interconnection point near the customer's premises. This interconnection point is a box on a telephone pole in a neighborhood or a wiring closet in an office building. From this interconnection point, the wires are routed to individual sites and end up at a network interface device (NID). The NID is the demarcation point between the telephone company side of the network and the inside wiring for your home or office.

Extending the reach of COs through DLCs

Digital Loop Carriers (DLCs) extend the reach of telecommunications services from a CO. DLCs are typically used in office parks and housing developments to minimize the need to run local loops over several miles to the CO servicing the area. Instead, local loops connect a cluster of homes or businesses to a remote terminal (RT), which in turn concentrates the traffic into a higher bandwidth for delivery to the CO, as shown in Figure 2-2. Remote terminals are typically green boxes located on a telephone pole or larger refrigerator-sized boxes on the ground.

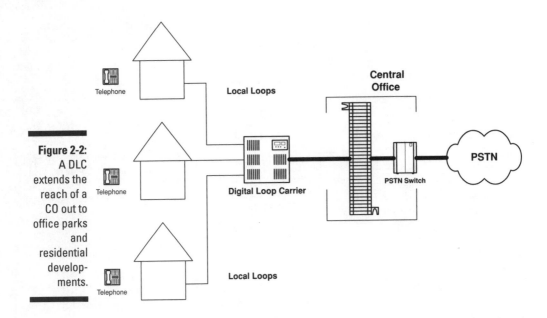

Figure 2-2:
A DLC
extends the
reach of a
CO out to
office parks
and
residential
develop-
ments.

By terminating loops at the DLC remote terminal, you reduce the effective length of the local loop and thus improve the reliability of the service. This architecture solved many problems for POTS, but it causes problems for the delivery of DSL services. The number one problem with DLCs is that the back (network) end is fiber, and DSL cannot travel over fiber. So the DSLAM must be in the DLC, and DLCs were not designed for this. So cards that offer DSLAM-like functionality must be made available that can slide into the DLCs. This has taken time, and only now are solutions to this problem being deployed. This is a critical availability issue because in some ILEC territories, DLCs can carry almost 50 percent of the traffic on average; however, DLCs cover 25 percent or so of the local loops in the United States. So, in order to make DSL a ubiquitous service, the DLC last mile problem has to be solved.

The big PSTN switch

The Public Switched Telephone Network (PSTN) uses digital switches to route telephone calls, with telephone numbers acting as a routing address system. *Switched services* are those in which the connection is made by some call control procedure, such as a person dialing a call to another telephone.

The PSTN uses circuit switching to transmit calls. A *circuit* is a physical path for the transmission of voice and data used only on an as-needed basis. The communications pathway remains fixed for the duration of the call and is not available to other users. A *circuit-switched connection* between two users becomes a fixed pathway through the network. At the end of the communications session, the pathway is terminated and is available to other users.

The Internet, the PSTN, and DSL

PSTN switches at the COs are sized according to probability. If everyone in the United States picked up their phone at the same time, only a small portion would be able to make calls. These switches are designed with the idea that a subscriber uses his or her phone line only a few times during the day. Therefore, instead of a 1-to-1 relationship between the customer's telephone line and the network, it's more like a 1-to-4 or 1-to-6 relationship.

The growth of the Internet has wreaked havoc on this probability environment because people are dialing up to their Internet provider and staying online for a long time — a lot longer than ever envisioned. As a result, it gives telephone subscribers in high-Internet subscriber areas poorer basic telephone service. For example, a switch designed to support 10,000 lines coming from customers' homes might support only 400 outbound lines from the CO for these customers to use. If 100 Internet subscribers stay online to surf the Internet, it reduces the availability of lines for everyone else by 25 percent — a large amount in the telephone industry.

This offloading of data traffic from PSTN switches to DSL lines connected to DSLAMs frees up PSTN switches for voice traffic. For ILECs, this is one of the big advantages of DSL.

Inside the DSL Network

The DSL network overlays and interacts with the PSTN network. DSL data communications uses key elements of the telecommunications network, most notably local loops and central offices. For ADSL connections with both data and POTS service on the same line, voice and data are separated. The POTS traffic goes to the PSTN switch to be handled as any telephone call. The data part goes to the DSLAM. The following sections explain the elements that make up the DSL system.

DSLAM: Center of the DSL universe

COs house DSLAMs (Digital Subscriber Line Access Multiplexers), which consolidate data traffic from individual DSL connections into large high-capacity backbone networks that connect to ISPs or corporate networks. Figure 2-3 shows the CopperEdge 200 concentrator, a DSLAM from Copper Mountain Networks. The DSLAM is a platform for DSL service aggregation as well as the gateway for IP routing, switching, and virtual private networking.

When a CLEC or an ILEC sets up DSLAM devices at a given CO, they make DSL service available to the area serviced by that CO. ILECs and CLECs choose COs that service large concentrations of potential DSL customers.

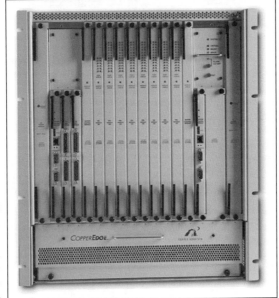

Figure 2-3:
CopperEdge
200 DSL
concen-
trator.

(Courtesy of Copper Mountain Networks, Inc.)

In the case of ADSL service, which supports POTS, DSLAMs split DSL's digital signal away from the analog voice service and route the voice traffic to the PSTN. This function is handled by a device called a POTS splitter. A *POTS splitter* is a device at the customer's premises that splits the data and voice there. G.Lite does not require a POTS splitter at the customer's premises but does require the use of filters on inside wires leading to POTS devices, such as telephones, fax machines, and modems.

Is your loop qualified?

DSL service requires a high-quality local loop with minimal interference. But copper wiring has been added to the PSTN since 1875, which means the quality of local loops across the PSTN varies. Even if a DSL provider says your premises are within the distance range for DSL service, you won't know whether you can get DSL service until a local loop is qualified.

Loop qualification is the process used to determine whether a specific copper pair will support DSL. The process starts with a *DSL screening,* in which you provide your 10-digit telephone number for the location where you want DSL service. (COs have their own addresses — the first six digits of your telephone number, which are the area code and exchange office code.) Typically, this process verifies whether the central office has a DSLAM installed. Adding the specific address of the premises where you want to install DSL service defines the approximate loop length between your premises and the CO. If the loop length is 14,000 feet or less, you'll be told that DSL is available in your area. The capability to deliver DSL service, however, can ultimately be determined only by an on-site testing of the local loop at your premises.

A variety of factors go into determining whether a local loop can support DSL, including the following:

- **DSLAM used at the CO.** Different DSLAMs supporting different DSL flavors have different capabilities.

- **Local loop wires.** Many are 24 or 26 AWG (American Wire Gauge). The AWG measures the thickness of the copper wiring. The thicker the wire, the less resistance it has for signals traveling over it. The thicker the copper wiring, the longer the distance that DSL service can be delivered.

- **Whether loading coils have been placed on the loop to improve voice quality on longer loops.** A loading coil is a metallic, doughnut-shaped device used to extend the reach of a local loop beyond 18,000 feet for POTS. Unfortunately, loading coils wreak havoc on DSL. If the local loop has any loading coils, they must be removed to use DSL.

- **Whether a bridge tap has been added to the local loop.** A bridge tap is an extension to a local loop generally used to attach a remote user to a central office without having to run a new pair of wires all the way back. Bridge taps branch off the main line. Bridge taps are fine for POTS but severely limit the speed of DSL service.

- **Spectrum incompatibility in a binder bundle.** The packaging of many copper wire pairs into a binder bundle has implications for the delivery of certain types of DSL service due to spectral interference (also called *crosstalk*), which happens when neighboring lines are corrupting each other.

Going the distance

The distance between your premises and the CO plays a critical role in determining whether DSL service can be delivered to you and at what speeds it can be delivered. Because of the physics of high-speed data communications, DSL service is distance sensitive. The maximum length of a local loop varies, depending on the DSL flavor.

Most DSL technologies have a distance limitation of 12,000 to 18,000 feet. By most estimates, 60 to 70 percent of the United States population live close enough to a central office to take advantage of the more popular DSL technologies. Of public switched networks, 20 percent can't handle DSL until the telephone companies remove devices that extend the distance that a signal can travel. Another large segment of the population supported by Digital Loop Carriers (DLCs) is also currently outside the loop in getting DSL service.

End of the Line

This section presents an overview of the DSL equipment and inside wiring that make up your end of the DSL connection.

Customer Premises Equipment (CPE)

In telecommunications lingo, the generic term used for DSL hardware used at your premises is *Customer Premises Equipment (CPE)*. Common forms of CPE for PC-based data communications are modems, bridges, and routers.

DSL CPE plays an integral role in the DSL service package because it defines what you can and can't do with your DSL connection. For example, a DSL provider might offer a high-speed DSL Internet access package that looks great — until you discover that the DSL CPE they're using restricts the service to a single computer. So there you are: Stuck with using only one PC with 1.54 Mbps downstream and 384 Kbps upstream, even though that rate could easily be used across a few computers. Controlling the number of PCs a connection can support is one way that the DSL provider can manage the bandwidth.

DSL CPE devices break down into single-user or multiuser solutions. Single-computer DSL CPE includes PCI adapter cards, Universal Serial Bus (USB) modems, and some bridges. The multiuser network environment has LAN modems (bridges), routers, and brouters.

The DSL CPE marketplace is chaotic. For starters, most DSL CPE devices are called modems, when in fact they're either bridges or routers. The difference in these technologies has a big effect on your DSL service. In some DSL offerings, the DSL CPE can end up restricting your DSL service and IP networking options. Choosing a DSL provider involves also taking a hard look at what CPE they're offering. ILECs are particularly notorious in using restrictive DSL CPE. See Chapter 5 for more details on DSL CPE.

The great divide

In the United States, telephone wiring responsibility is divided between the telephone company, which is responsible for local loop wiring to your doorstep, and you, the customer, who is responsible for any inside wiring. In homes, a network interface device (NID) usually separates the telephone company's wiring and your premises wiring. Figure 2-4 shows a network interface device. The NID allows the telephone company installer to bring the line to your home. From this box, you or the telephone company installer attaches the twisted-pair wiring that leads to the specific location of your CPE.

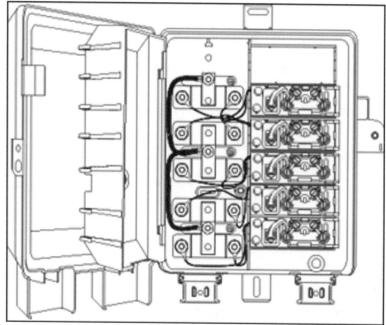

Figure 2-4:
A Siecor
network
interface
device
(NID).

If your premises are in an older structure, you might have a *protector block* instead of an NID. The FCC prohibits customers from working at the protector block, so the telephone company must install an NID for you to get DSL service. The FCC also has the *12-inch rule,* which states that if you have a protector block, the NID must be located within 12 inches of it.

Getting wired for DSL

POTS and DSL service require the same physical wiring. Depending on the DSL flavor installed, you might need to get a new line installed or you might be able to convert an existing telephone line to DSL.

Standard telephone wiring consists of single 4-wire cable, which allows for two lines. For residences outside metropolitan areas, the number of telephone lines may be restricted to two.

For your inside wiring from the NID to the location where you plan to use your DSL CPE, the installer of your DSL line (ILEC or CLEC) may do the inside wiring. DSL service is handled differently than traditional wiring because the installation of the service typically involves a certain specified amount of additional inside wiring to bring the line to the DSL CPE location. To what extent inside wiring is handled by a DSL provider is spelled out in your DSL service.

Most ADSL deployments enable you to share the same line for both POTS and DSL service through the use of a POTS splitter. A *splitter,* which is installed by the DSL provider, separates the POTS channel from the high-speed DSL channel. ADSL typically uses a POTS splitter that connects to the same line as the DSL CPE. You then plug your POTS device into the POTS splitter. G.Lite eliminates the POTS splitter but still requires the installation of small filter devices to the line next to every POTS device (telephone, fax, or modem) sharing the G.Lite line, as shown in Figure 2-5.

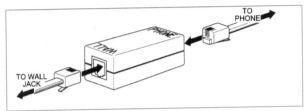

Figure 2-5:
The EZ-DSL
Microfilter.

Know your connectors

At the end of your wiring are connectors. For DSL service, the connector jack used for your DSL wiring can be either an RJ-11 or RJ-45 modular connector, which snaps into a jack. You can plug an RJ-11 connector into an RJ-45 jack. Figure 2-6 shows the RJ-11 and RJ-45 connectors.

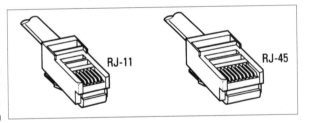

Figure 2-6:
The 4-wire
RJ-11 and
8-wire RJ-45
modular
connectors.

CLECs and DSL Service

DSL CLECs are commonly referred to as data CLECs or packet CLECs. Whatever they call themselves, these CLECs are conversant in the language of the local loop but also grounded in the world of the Internet and IP networking.

CLECs are leading the charge in the delivery of DSL services in many areas. Their data-only approach makes them more agile in rolling out DSL services than ILECs, which support the old, dual-purpose, voice and data infrastructure.

The two leading data CLECs are NorthPoint Communications and Covad Communications. NorthPoint Communications focuses exclusively on delivering business-class data communications services at affordable prices. NorthPoint markets its DSL services through Internet service providers in most major metropolitan markets. NorthPoint functions as the WAN portion of your Internet connection and offers SDSL (Symmetric DSL) service with a choice of data rates — 160 Kbps, 416 Kbps, 784 Kbps, 1.04 Mbps, and 1.54 Mbps — and at prices well below costly ILEC T1 and fractional T1 services.

Covad Communications also aggressively prices SDSL, ADSL, and IDSL services through ISPs in most large metropolitan areas. Many of Covad's ISP partners are the same as NorthPoint's partners. Other CLECs offer DSL service on a local or regional level.

Some ISPs have also become CLECs to deliver DSL Internet connections. In the Boston area, HarvardNet is a local ISP acting as a CLEC offering DSL service. The leading national DSL data CLECs are Covad Communications, NorthPoint Communications, and Rhythms NetConnections.

How a data CLEC operates

Within each CO, the CLEC maintains DSLAMs in racks that connect to unbundled local loops supplied by the ILEC. The CLEC works with the ILEC to get the twisted-pair wiring delivered to the network interface device at the customer's premises. After the ILEC installs the twisted-pair, the CLEC installer does the inside wiring, tests the line, and sets up the DSL CPE.

The CLEC maintains a backbone network center that acts as the central hub for data traffic coming from DSLAMs in multiple central offices throughout a metropolitan area. The ISP taps into the CLEC's central hub through a single fat pipe to move IP traffic.

The CLEC defines which areas have DSL service available by determining which COs get DSLAMs. The CLEC manages all CPE installations, maintains the local loop and colocation sites, and monitors and manages the internetwork connecting all COs to the central hub.

CLECs, ISPs, and DSL service

The relationship between most CLECs offering DSL service and the ISP is a partnership, with the CLEC acting as the silent partner. The CLEC makes the DSL service available to the ISP on a CO-by-CO basis. The ISP typically buys the circuit for a fixed cost based on the speed of the connection. The ISP provides the customer interface to the combined DSL data communications link and Internet access service. The ISP offers the IP networking services, including Domain Name Server (DNS) services, mail services, Web hosting, and any other customization services.

The long and winding road to DSL service

Although the Telecommunications Act requires that ILECs open up their markets to CLECs, the journey to get to the point of delivering DSL service is down a long and winding road. A CLEC must file to become a telecommunications provider in any state where it wants to offer DSL service. The CLEC must then negotiate a contract with the ILEC, which can be extremely difficult if the ILEC is uncooperative. After a deal is made, the CLEC must begin the process of building its infrastructure at each CO from which it wants to offer DSL service.

The ISP ultimately creates the actual DSL service package and pricing. The cost of the CLEC's DSL service, commonly called the *DSL circuit,* is typically included in a single bill from the ISP. Although CLECs typically install the link at the customer's premises, the ISP configures the DSL CPE, which is shipped to the premises before the DSL service is installed.

ILECs and DSL Service

Most ILECs are deploying only ADSL and (soon) G.Lite because they support high-speed data communications and POTS service on the same line. ILEC DSL service offerings are typically broken down by speed, with the lower speed offerings targeted at the home market and the higher speed offerings targeted at the business market.

Chapter 9 provides a detailed listing of ILEC DSL offerings.

ILECs are offering DSL service using two methods. The first method is one in which the ILEC partners with an ISP who then resells the ILEC's services. This means that you can get ILEC-resold DSL service through ISPs. The DSL transport charges are typically (but not always) added to the single bill from the ISP. The second method for offering DSL service is as a bundled service with Internet access, provided by both the ILEC and its in-house ISP, such as PacBell.net.

If you get a new line, you might be required to sign up for POTS service as well, because many ILEC DSL implementations are ADSL that supports POTS.

Chapter 3

The Zen of DSL

In This Chapter

▶ Grasping the fundamentals of DSL data communications

▶ Tasting the different flavors of DSL

▶ Checking out the DSL standards scene

*B*y understanding the basics of bandwidth and data communications, you'll build the foundation you need to understand the lingo and navigate through the maze of DSL technologies. This chapter provides an overview of the DSL technology that makes it possible to use copper telephone lines to deliver high-speed data communications.

DSL Data Communications 101

DSL service takes you into the world of digital data communications. Terms such as *digital, bandwidth, asymmetric,* and *symmetric* are spoken in the everyday world of DSL. This section explains the essential concepts and terminology associated with DSL data communications. Grasping these buzz words will allow you to speak the local language to better navigate the world of DSL service.

The nature of digital communications

Digital communication is the exchange of information in binary form. Unlike an analog signal, a digital signal doesn't use continuous waves to transmit information. Instead, a digital signal transmits information using two discrete signals, or on and off states of electrical current. All computer data can be communicated through patterns of these electrical pulses. Digital communications support high-speed data communications and are more reliable than analog communications. Because DSL is a digital form of data communications, your Internet connection is more reliable.

Digital data may be represented in many ways, so a special interface device must convert digital content from the computer to a form of digital signaling used for the data communications link. For this digital conversion, we use a functional device generically called a CSU/DSU (Channel Service Unit/Data Service Unit). The term *functional device* refers to the activities performed by a specific device. The channel service unit (CSU) and data service unit (DSU) are actually two separate functional devices that are usually combined in the same device. The CSU terminates the digital data communications line. The DSU converts the digital signals coming from a computer networking device (such as a bridge or a router) into a digital signal understood by the data communications link.

The conversion of data is the reason many DSL CPE products are called DSL modems. The term *DSL modem,* however, is more a marketing term than a technical term. DSL modems do not perform the modulation/demodulation performed in analog modems. The term DSL modem is more of a metaphor to describe its modem-like functionality in making the connection to the Internet through the DSL WAN.

Frequencies and bandwidth

The capacity of a data communications line is based on the medium and the frequency range used by a given data transmission technology. *Media* is the stuff on which voice and data transmissions are carried, which in the case of DSL is the twisted-pair copper wires that make up the local loop. *Bandwidth* refers to capacity as defined by the combination of the media used and the frequency spectrum that the media can support. *Frequency* equals the number of complete cycles of electrical current occurring in one second, which is measured in Hertz, or cycles per second. The higher the frequency, or Hertz, the greater the capacity, or bandwidth.

The use of higher frequencies in DSL to support higher data communications results in shorter local loop reach. This is because high-frequency signals transmitted over copper loops dissipate energy faster than lower frequency signals. The electrical properties of copper wiring create resistance and interference problems with data transmissions. This forms the basis of the inherent limitations of DSL service based on the distance between a particular office site and the central office servicing a given area.

The concept of frequency is important in understanding how the same telephone line can be used for analog voice service and yet also support higher bandwidth connections. For example, ADSL can use the same line as POTS because ADSL exploits a higher frequency than that used for voice service. POTS carries voice communications in the voice frequency range of 300 to 3,300 Hz. ADSL uses a much broader range of frequencies that range from 4 kHz to 1.1 MHz.

Changing channels

The term *channel* in communications is any pathway used for data transmission. A communications channel is a logical pathway defined separately from the media that delivers the data communications. Each channel is an independent unit that can transfer data concurrently with other channels. Many digital communications technologies, including DSL, use multiple channels. For example, an ADSL connection includes two channels, one for the data and the other for POTS. Each operates separately.

Signaling and modulation

Signaling is the process whereby an electrical signal is transmitted over a medium for communications. To transmit data across the twisted-pair copper wiring between your premises and the CO, some method of signaling must be used. High-frequency signals transmitted over copper loops dissipate energy faster than lower frequency signals. Modulation techniques minimize the loss of electrical energy as it passes over a copper wire by reducing the frequency, which in turn extends the local loop reach.

Modulation is based on different line code schemes. Line coding techniques are an important factor in the deployment of DSL. Some line codes are spectrally incompatible with other line coding. If two line codes are incompatible, their frequencies spill over into adjacent wire pairs and interfere with the signal.

The DSL realm has three leading line codes: CAP (Carrierless Amplitude and Phase), DMT (Discrete MultiTone), and 2B1Q (Two Binary, One Quaternary). 2B1Q and DMT are the most spectrally compatible. Table 3-1 lists the DSL flavors and their respective line code schemes.

Table 3-1	DSL Flavors and Their Line Codes
DSL Flavor	**Line Code**
ADSL	CAP and DMT
IDSL	2B1Q
SDSL	CAP and 2B1Q
RADSL	CAP and DMT
HDSL	CAP and 2B1Q

Asymmetric or symmetric

In many forms of data communications (including DSL), one data channel supports a larger data communications capacity than the other channel. This unequal form of data communications is referred to as *asymmetric,* which means data traveling in one direction moves faster because of a higher capacity than data moving in the opposite direction. In the case of Internet access, an asymmetric connection means data coming from the Internet to your computer travels at a much higher rate than data going from your computer to the Internet.

In *symmetric* data communications, both channels have equal capacity for uploading and downloading data. The DSL family of technologies has both symmetric and asymmetric versions. Note, though, that asymmetric versions of DSL can be packaged as symmetric services by matching the downstream speed with the maximum upstream capability of the DSL service.

Businesses typically want symmetric data traffic to enable equally needed data speeds for running Web and e-mail servers as well as for upstream file transfers, video conferencing, and IP telephony. Symmetric service offered by most CLECs is Symmetric DSL, which fits the need for equal speed.

Consumers can typically work well with asymmetric service because they're browsing the Web and downloading stuff from the Internet but don't move upstream much beyond text e-mail messages. Because ILECs are targeting consumers, they're using Asymmetric DSL offerings.

Asymmetric service may be sufficient for many consumers, but be careful about some ADSL service offerings that restrict the speed of data going from your computer to the Internet. One ILEC, Bell Atlantic, supports only 90 Kbps, which precludes you from using some cool stuff such as video conferencing and other two-way, higher bandwidth Internet applications.

Data swimming upstream or downstream

The carrying capacity of a communications link is measured by the two directions that data moves: upstream and downstream. *Downstream* refers to data traffic moving from the Internet or another remote network to your computer or network. *Upstream* refers to data moving from your computer or network to the Internet or another remote network. If you're uploading a file to an FTP server on the Internet, for example, the upstream capacity of your DSL connection determines how fast the file will be uploaded. Likewise, if you download a file from an FTP server, the downstream capacity of your connection determines how fast the file is transferred. For DSL service, you'll see references to downstream and upstream speeds as, for example, 1.5 Mbps/384 Kbps, which means that the service offers 1.5 Mbps downstream and 384 Kbps upstream.

Throughput (bandwidth in the real world)

Bandits of bandwidth lurk on any data communications link. These bandits come from a variety of factors that affect the true throughput of a connection. *Throughput* is an overall measure of a communication link's performance in terms of its real-world speed. Think of throughput as a mileage rating for a new car. Claims made by bandwidth providers are like gasoline mileage claims made before the government standardized the methods for setting such claims.

Having the fastest link to the Internet is only half the story of what your real throughput will be. Remember, the Internet itself is one giant, shared network. It suffers from rush hour traffic just as highways do. With heavy network traffic and a crowded Web server, your speed can dramatically slow down.

A number of factors can affect your connection and make the throughput figure lower:

- ✔ **Internet traffic volume between your computer or network and the remote server.** If the Internet is experiencing a heavy volume of data traffic, the entire system slows down.

- ✔ **The speed of the server handling your request.** If the server is busy, it will slow down in terms of delivering your requested data. With faster DSL connections, you'll begin to notice the slower Web servers, something you may have never noticed when you used an analog modem.

- ✔ **The quality of your DSL connection to the ISP.** A noisy telephone line can require data to be re-sent more frequently.

- ✔ **The backbone used by the ISP to connect to the Internet.** If an ISP is oversubscribing DSL customers by not providing enough data communications capacity to the Internet, your throughput will deteriorate.

Several sites on the Internet, including the Internet Weather Report (www.internetweather.com) and the Internet Traffic Report (www.internettrafficreport.com), monitor Internet traffic patterns to help you determine the source of your sluggish Internet connection. Also available for monitoring your Internet connections are software tools such as the Visual Trace Route Utility (www.visualroute.com) and NetMedic (www.vitalsigns.com). More on these throughput analysis tools in Chapter 8.

The Many Flavors of DSL

Life would be easier if there was just one form of DSL. But who said life was easy? Each different DSL form, or flavor, has different capabilities and configurations. In addition, different DSL providers offer different DSL flavors. Understanding the attributes of each DSL flavor will go a long way in helping you analyze your DSL options.

ADSL

Asymmetric Digital Subscriber Line (ADSL) delivers a range of data communications speeds. Downstream speeds can reach up to 8 Mbps, and upstream speeds can reach up to 1 Mbps. ADSL can be packaged as a symmetric service up to full upstream capabilities. ADSL is being delivered using both CAP and DMT modulation. ADSL is commonly referred to as ADSL Heavy because of its high-speed capabilities and because of a new version of ADSL called G.Lite.

ADSL distinguishes itself from other forms of DSL by its use of a POTS splitter. The splitter allows existing analog voice and data services to coexist on the same line as the one used for the high-speed data service. POTS splitters are filters used at each end of the local loop to split the data traffic between low-frequency voice communications and high-frequency data communications.

Because ADSL shares the same line used for POTS service, you can convert an existing POTS line to an ADSL line, which makes it an attractive option for the residential and small business market. However, the line coding used for ADSL can cause crosstalk problems in cable bundles. This problem may hamper the availability of ADSL in many areas, especially areas with large numbers of T1 lines installed (although it is far less an issue with G.Lite or UADSL).

Like all other DSL members, ADSL is distance sensitive. The longer the distance between your premises and the CO, the lower your speed options. Table 3-2 shows the maximum downstream speeds based on selected distances. Keep in mind that actual distance and speed parameters depend on other technical factors.

Table 3-2	ADSL Speeds at Different Distances
Distance of Your Premises from the CO	*Downstream Speed*
Up to 9,000 ft	8 Mbps
Up to 12,000 ft	6 Mbps
Up to 16,000 ft	2 Mbps
Up to 18,000 ft	1.5 Mbps

UADSL (G.Lite)

Universal ADSL (UADSL), which is based on the G.Lite standard, is intended primarily for the consumer Internet access market and can deliver up to 1.5 Mbps downstream and up to 512 Kbps upstream. The G.Lite standard offers these benefits to DSL consumers:

- ✓ **Reduced installation expense.** UADSL supports both high-speed data and POTS service over the same line. But unlike ADSL, UADSL doesn't require a POTS splitter; this reduces installation costs.

- ✓ **Competitively priced DSL modems.** G.Lite standardization enables a competitive CPE vendor environment that promises to dramatically drop the price of DSL CPE and make them available in the computer retail channel.

- ✓ **Greater accessibility.** G.Lite operates at lower frequencies than ADSL. This allows G.Lite to transmit data over longer distances, which enables more people to have access to it. G.Lite supports data communications up to a distance of 18,000 feet between the customer's premises and the CO.

Although G.Lite is referred to as splitterless ADSL, it does use a distributed splitter approach based on simple inline low-pass filters (LPFs) placed at each phone in the customer premises combined with a high-pass filter (HPF) built into the ADSL modem. Figure 3-1 shows a typical microfilter. These microfilters are needed because POTS devices (telephones, faxes, and modems) can cause interference for a UADSL connection. Figure 3-2 shows how a G.Lite (UADSL) configuration would be configured using LPFs.

Figure 3-1:
A typical LPF microfilter used with a G.Lite (UADSL) installation.

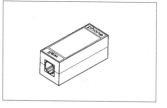

Figure 3-2:
A G.Lite configuration using LPFs at each POTS device.

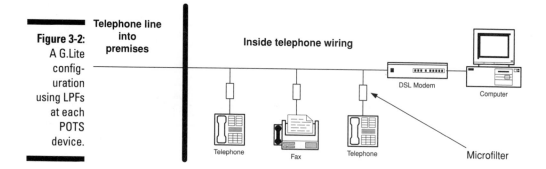

G.Lite is based on the ITU's G.992.1 and G.992.2 standards. G.Lite uses DMT line coding technology, which is based on the ANSI standard T1.413 Issue 2 DMT standard. One unknown about G.Lite is how well it will behave in binder groups and how well it will filter out interference from POTS devices in a customer's premises.

SDSL

Symmetric Digital Subscriber Line (SDSL), as its name implies, is a symmetric service that can deliver variable high-speed data communication speeds up to 1.54 Mbps. SDSL can be packaged in a range of bandwidth configurations that include 144K, 160K, 192K, 384K, 416K, 768K, 1 Mbps, and 1.5 Mbps.

SDSL, unlike ADSL, doesn't support the use of analog on the same line. SDSL uses 2B1Q line coding, which is widely used by the telephone companies for such services as ISDN and T1 lines. SDSL and HDSL are related members of the DSL family. HDSL is a proven DSL technology that has been used in T1 lines for years.

Because of its high data rates, ease of deployment (made possible by its spectral compatibility), and symmetric transmission, the business market finds SDSL attractive. SDSL is being aggressively deployed by the leading CLECs, such as NorthPoint Communications and Covad Communications. Table 3-3 shows the speed parameters for SDSL service based on the distance between the customer's premises and the CO. However, other technical factors come into play in determining the actual speed available at your premises.

Table 3-3	SDSL Distance and Speed Parameters
Distance of Your Premises from the CO	*Speed*
Up to 9,500 ft	1.5 Mbps
Up to 12,500 ft	1.04 Mbps
Up to 14,900 ft	784 Kbps
Up to 18,000 ft	416 Kbps
Up to 19,000 ft	320 Kbps
Up to 20,000 ft	208 Kbps
Up to 22,770 ft	160 Kbps

RADSL

Rate-adaptive Asymmetric Digital Subscriber Line (RADSL) is the rate adaptive variation of ADSL. Most ADSL deployments are really RADSL versions. RADSL can support symmetric communications up to 1 Mbps and asymmetric downstream speeds of 1.5 Mbps to 7 Mbps. RADSL operates within the same transmission rates as ADSL but adjusts dynamically to varying lengths and quality of twisted-pair local access lines. This is the rate-adaptive function of RADSL. RADSL also supports POTS.

IDSL

ISDN Digital Subscriber Line (IDSL) is the always-on cousin of dial-up ISDN (Integrated Services Digital Network). IDSL delivers a symmetric 144 Kbps of bandwidth, which is 16 Kbps more than the dial-up version of ISDN. This 16-Kbps difference comes from the elimination of the two 8-Kbps channels used in ISDN for communicating with the PSTN switch. IDSL uses the 2B1Q line coding scheme and doesn't support analog as does ISDN.

IDSL is attractive because it has a much larger range than the other DSL flavors, reaching 26,000 feet or more. ISDL offers a more affordable alternative for many dial-up ISDN users facing high usage costs. Another advantage to making the ISDN-to-IDSL conversion is that you can use existing ISDN CPE with IDSL.

The downside of IDSL is that it's not scalable, which means you can't upgrade your speed. ILECs and CLECs are both offering IDSL services to address the larger potential market than available with their other DSL services alone.

HDSL and HDSL-2

High-bit-rate Digital Subscriber Line (HDSL) is the most mature DSL technology. It's been used by ILECs for years as the basis for T1 lines. HDSL provides symmetric data communications of up to 1.54 Mbps. HDSL requires the use of two pairs (four wires) instead of the standard single pair used for other DSL flavors and doesn't support POTS. HDSL uses DMT, CAP, and 2B1Q line coding. T1 service using HDSL technology has become very profitable for ILECs, but they're facing competition from CLEC SDSL offerings. SDSL is a rate-adaptive version of HDSL.

The ITU has approved a new generation of HDSL called *HDSL-2* that offers several enhancements over its predecessor. One of the most important is that HDSL-2 requires only a single twisted-pair local loop instead of the two pairs required for HDSL. HDSL-2 delivers its data communications service over a single copper-wire pair to a distance of up to 18,000 feet without the use of repeaters. HDSL-2 promises to lower the cost of T1 services.

VDSL

Very-high-bit-rate Digital Subscriber Line (VDSL) is the ultra-high-speed DSL flavor. VDSL is an asymmetric technology that can be used only by premises in close proximity to a central office. The closer the customer is to the CO, the higher the speed capabilities. For example, at 1,000 feet from the CO, the downstream rate can reach up to 52 Mbps. At 3,000 feet, a subscriber can attain data rates up to 26 Mbps, and at about 5,000 feet, data rates can reach up to 13 Mbps. VDSL can provide asymmetric data and POTS transmissions on a single twisted pair.

MVLDSL

Multiple Virtual Lines DSL (MVLDSL) is a proprietary version of DSL developed by Paradyne. Up to eight MVLDSL modems can be connected to a single POTS wire pair, enabling each to have a bandwidth of up to 768 Kbps downstream and upstream. Data can be transmitted up to 30,000 feet. MVLDSL is a splitterless technology, so service technicians are not required at the customer's premises.

Within the customer site, MVLDSL modems can be connected to phone jacks, and then PCs can send data between modems in the same building as if they were on a LAN. MVLDSL is considered a very good technology that works where a lot of other DSL technologies don't.

DSL Standards and Forums

Standards are agreed principles of protocol set by committees working with various industry and international standards organizations. The standards process works through consensus. Experts from across an industry meet, debate, and share options and research to arrive at a standard. Following are the big benefits consumers gain from standards:

- ✔ **Standards fuel competition** among CPE manufacturers and data communication service providers.

- ✔ **Standards usually create a solution that is greater than the sum of its parts.** The process of different experts from a variety of companies working together and discussing the technology leads to a solution significantly better than one company could develop on its own. Note, however, that standards are compromises of great competing working solutions.

✔ **Standards lead to interoperability among devices,** so consumers can buy products from multiple vendors.

✔ **Standards create a level playing field** that allow service providers and product vendors to offer DSL CPE and services at the lowest possible prices due to competition.

Know your DSL standards organizations

Two major standards bodies influence DSL:

✔ The **International Telephone Union (ITU)** is an agency of the United Nations that is the primary source of telecommunications standards. You can check out the ITU on the Web at `www.itu.int`.

✔ The **American National Standards Institute (ANSI)** is the primary standards setting body in the United States. ANSI plays a significant role in the development of DSL standards in the United States. ANSI T1.413 is the standard for DMT in the United States. You can check out ANSI on the Web at `www.ansi.org`.

Two leading DSL industry organizations are involved in shaping proposed standards. These organizations act as forums in which DSL industry players work together to develop a proposal for the standards bodies. The two leading DSL industry forums follow:

✔ The **ADSL Forum** is one of the leading industry organizations working to develop consensus on DSL standards. The ADSL Forum represents most telecommunications and computer companies involved with DSL services. You can check out the ADSL Forum on the Web at `www.adsl.com`.

✔ The **Universal ADSL Working Group (UAWG)**, which is made up of telecommunications and computer companies, has focused on developing G.Lite. The UAWG was created as a temporary group, specifically to expedite the recommendation of a set of technical requirements to the ITU, facilitate vendor interoperability testing, and aid in the removal of potential barriers towards mass deployment of ADSL services. After these tasks have been completed, the UAWG will hand off the torch to permanent organizations such as the ADSL Forum and the ITU. You can check out the UAWG on the Web at `www.uawg.com`.

The current state of DSL standards

The current state of DSL standards is one of steady progress. Although universal DSL CPE interoperability isn't in place, some big standards developments have occurred. The G.Lite standard and HDSL-2 are two important standards that will have an effect on consumer DSL and business DSL service, respectively. Other standards related to technical background issues for DSL have also been established, such as G.994.1. This standard deals with the handshakes, test procedures, and behind-the-scenes technical issues used by DSL modems. G.994.1 is specified for ADSL but will be adopted for HDSL and HDSL-2.

For other forms of DSL, such as SDSL, the lack of standards means that CPE and DSLAMs manufactured by different vendors do not automatically interoperate. Over time, however, standards will emerge to enable DSL customers to buy their own CPE from any vendor. In the meantime, DSL providers control the deployment of CPE to ensure that it works with their DSLAM equipment. Many DSLAM vendors work with multiple CPE vendors to give users some choices in their hardware.

Chapter 4

All Nodes Lead to TCP/IP

*T*CP/IP is the universal language for computer communications on the Internet. When you use DSL to connect to the Internet, you must configure your LAN for Internet access through TCP/IP networking. The type of IP addressing you use as part of your DSL Internet service plays a big role in determining what you can and can't do with your Internet connection.

Understanding your TCP/IP options empowers you to make good choices in configuring your LAN. This chapter explains the fundamentals of TCP/IP networking as a backdrop for configuring your LAN for a DSL connection.

TCP/IP and Your DSL Connection

A *protocol* is a set of rules used to allow interoperability among different systems. TCP/IP, or Transmission Control Protocol/Internet Protocol, is the lingua franca of the Internet. TCP/IP is actually a combination of two key protocols. *Transmission Control Protocol (TCP)* is the transmission layer of the protocol and serves to ensure reliable, verifiable data exchange between hosts on the network. (A *host* is any computer or other device connected to a network.) TCP breaks data into packets, wrapping each with the information needed to route it to its destination, and then reassembles the packets at the receiving end of the communications link.

TCP puts in the packet a *header* that provides the information needed to get the data to its destination and then passes the packet to the *Internet Protocol (IP)*, which is the networking layer of TCP/IP. IP inserts its own header in the packet and then moves the data from point A to point B through a process called *routing*.

IP is referred to as *connectionless* because it does not swap control information (handshaking information) before establishing an end-to-end connection and starting a data transmission. IP relies on TCP to determine whether the data arrived successfully at its destination and to retransmit the data if it did not. IP's only job is to route the data to its destination. IP inserts its own header in the packet when it's received from TCP.

The Internet Protocol (IP) is responsible for basic network connectivity and is at the heart of how DSL CPE works. The Internet Protocol (IP) is responsible for basic network connectivity. The core of IP works with Internet addresses. Every computer on a TCP/IP network must have a numeric address. Although IP can take care of addressing, it can't do everything to make sure that your data gets to where it's going correctly and in one piece. IP doesn't know or care when a packet gets lost or doesn't arrive. So you need other protocols to ensure that no packets are lost and that the packets are in the right order.

The five layers of TCP/IP

TCP/IP takes a layered protocol approach to networking, which means each protocol is independent of the others but all protocols work together to enable TCP/IP networking. TCP/IP loosely follows the Open Systems Interconnection (OSI) Reference Model, which defines the layers and protocols that networks must contain to control interactions between computers. The OSI model defines the framework for implementing protocols in seven layers: application, presentation, session, transport, network, data link, and physical.

When information passes from computer to computer through a protocol, control of the data passes from one layer to the next, starting at the application layer and proceeding through to the physical layer. The information then proceeds to the bottom layer of the next system and up the hierarchy in that second system.

The TCP/IP protocol consists of five layers that perform the functions of the OSI model's seven layers.

Each layer adds its own header and trailer data to the basic data packet and encapsulates the data from the layer above. On the receiving end, this header information is stripped, one layer at a time, until the data arrives at its final destination.

The five layers of TCP/IP are described in Table 4-1. The Application layer combines the OSI's session, presentation, and applications layers, which explains the five layers of TCP/IP versus the seven layers of the OSI Reference Model.

Table 4-1	The Five Layers of TCP/IP
Layer	*What It Is*
Physical (Layer 1)	The pure hardware layer, including the NICs, cabling, and any other piece of hardware used on the network.
Data link (Layer 2)	The layer that splits your data into packets to be sent across the connection medium. It interfaces with the physical layer, which is the hardware and network medium.
Internet (Layer 3)	The layer where the IP (Internet Protocol) fits into the equation. This layer gets packets from the data link layer and sends them to the correct network address.
Transport (Layer 4)	The layer that makes sure your packets have no errors and that all packets arrived and are in the correct order. This layer transports data through TCP and passes it to the Internet layer.
Application (Layer 5)	The layer where you do your work, such as sending e-mail or requesting a Web document from a Web server. At this layer, you're working with protocols that form the basis of TCP/IP applications, such as HTTP for the Web and FTP for file transfers.

Why do you need to understand the OSI and TCP/IP layers? Different DSL CPE work at different layers of the OSI and TCP/IP network models. You'll often see references to different DSL CPE working at different levels. This difference defines the two leading DSL CPE devices: bridges and routers. A DSL modem (a bridge) works at the OSI and TCP/IP layers 1 and 2, but DSL routers go up to layer 4. The higher the layer at which a DSL CPE works, the more sophisticated its capabilities. Chapter 5 goes into more detail on DSL CPE options.

The life of packets

TCP/IP is based on a technology known as packet-based networking. In a *packet network*, data travels across a network in independent and variable-sized units that can be routed over different network paths to reach the ultimate destination.

The life of a packet (also called a *datagram*) begins when an application, such as a Web browser, creates it. As each packet travels down the sending computer's layers, it picks up additional control and formatting information in a header to ensure its delivery to the destination computer. Each router on the network that encounters a packet examines the header to determine whether the packet is intended for its local network. If not, the packet is passed on in a direction closer to the ultimate destination. This process of getting your data from point A to point B through the TCP/IP network is called *routing*. When the packet reaches the destination computer, the header is read and stripped out as the packet moves up through each layer to the application.

The benefit of the packet-switched network is that packets in a message do not have to travel to the destination along the same route. Packets can travel many different routes, but all end up at the same destination, where they are reassembled in the order originally intended. This independent routing of packets over a network allows data to be transmitted even if parts of the network are disrupted.

Gateway to the TCP/IP universe

Each host on a TCP/IP network has a *gateway*, which is the off-ramp for packets not destined for other hosts on the local network. This forms the core of IP networking. A *router* embodies the gateway function, and both terms are used interchangeably. Each gateway has a defined set of routing tables that tell it the route to specific destinations.

Because a gateway can't know the location of every IP address, it has its own predetermined gateway to which it forwards all packets of unknown destination. This forwarding, or routing, continues until the packet reaches its destination. The entire path to the destination is known as the *route*. The number of gateways in a transmission path between two hosts on the Internet is referred to as the number of *hops*.

A router allows data to be routed to different networks based on packet address and protocol information associated with the data. Routers read the data passing through them and decide where the data is sent. This decision-making functionality, called *filtering*, allows or disallows certain source IP ranges, or protocols, from either entering or passing. With filtering, a router can help protect your network from unwanted intrusion and prevent selected local network traffic from leaving your local network. Filtering is not perfect, however, and can be hacked. Chapter 6 explains Internet security issues.

The gateway function is built into DSL CPE devices that connect a LAN to the Internet. Whether you connect a single computer or a LAN to a DSL device, that device is the gateway to the outside world. The fundamentals of internetworking devices are similar regardless of the flavor of DSL you're using.

Any port in a storm

On a TCP/IP network, data travels from a port on the sending computer to a port on the receiving computer. A *port* is an address that identifies the application associated with the data. Table 4-2 lists common Internet application port addresses.

Table 4-2	Common TCP/IP Port Numbers
Internet Service	*Port*
World Wide Web	80
FTP	21
SMTP	25
News	144

Ports are typically transparent to network users because many ports are universally assigned to specific TCP/IP services. For example, Web servers normally run on port 80, and FTP servers run on port 21. This means when users connect to a Web server, their Web browsers automatically connect to port 80. If a Web server is using a different port number, a user who doesn't know the unique TCP/IP port can't access it. An example of the URL for a nonstandard Web site is `http://www.angell.com:8001`, where `8001` is the port number.

Typically, port numbers above 255 are reserved for private use of the local machine, and numbers below 255 are defined as defaults for a variety of universal TCP/IP applications and services. Port numbers can be created also using any port number above 255. TCP/IP doesn't see the specific application. It sees only the numbers — the Internet address of the host that provides the service and the port number through which the application intends to communicate.

Ports come into play in several aspects of your DSL-to-Internet connection. If you're running a DSL router, proxy server, or firewall, ports play a big role in defining what data can travel between your LAN and the Internet. Chapter 6 goes into more detail on firewalls and proxy servers.

Creating Identity with IP Addresses

Within any networking protocol, there must be a way to identify individual computers or other networked devices. TCP/IP is no exception and includes an addressing scheme that pervades the Internet and intranets. Think of Internet Protocol (IP) addresses as the unique telephone numbers for specific computers or other network devices on any TCP/IP network.

An *IP address* is a software-based numeric identifier assigned to each machine on an IP network. Each computer or other network device that uses TCP/IP is distinguished from others on the network by this unique IP address. IPv4 IP addresses are 32-bit addresses represented as decimal values between 0 and 255.

An IP address is organized into four groups of 8-bit numbers separated by periods, or dots, such as 199.232.255.113. The 199.232.255 is the network or subnetwork, and the .113 is the host address. The *network address* uniquely identifies each network. Every machine on the same network shares that network address as part of its IP address. The *node address,* or *host address,* is assigned to each machine on a network.

IP addresses are very difficult for humans to remember, so easier-to-remember *domain names* are mapped to each IP address. In that way, we humans can refer to a specific computer or device by its domain name rather than its IP address. The Domain Name System (DNS), which is explained later (in the "What's In a Name Anyway?" section), provides the friendly text interface to IP addresses.

An essential piece of your DSL service is the type of IP address configuration: a dynamic IP address Internet service or a static IP address Internet service. The key distinction here is that using a *dynamic IP* type of Internet access account means that your PC or LAN is invisible to the Internet. This configuration is targeted at the consumer or teleworker who doesn't plan to run any type of Internet server or Net voice or video conferencing applications.

Static IP addresses are recognized and routable on the Internet. Static IP addressing is typically used by businesses and power users to enable them to run Internet servers and such applications as Net voice and video conferencing. Static IP address Internet access accounts cost more because you must lease IP addresses from the ISP.

IP addresses have class and no class

Traditionally, IP addresses were assigned to networks using three classifications based on size: Class A, Class B, and Class C. This breakdown of IP addresses was simply a way of allocating addresses among the different networks that access the Internet:

- **Class A networks** are the El Grande of IP networks. Only 126 Class A addresses are possible in the world, and each Class A network can have in excess of 16 million computers in its individual networks.

- **Class B networks** can have up to approximately 65,000 workstations on the network. Approximately 16,000 Class B networks are in the world.

- **Class C networks** can have up to 254 workstations on the network. Several million Class C networks are possible.

IPv6: The next generation

The current version of IP addressing is IPv4 (for version 4). Over the next ten years, at least, a new generation of IP, officially named IPv6, will be phased in. Aiding the gradual transition is the fact that IPv4 and IPv6 can coexist. One of the main reasons behind IPv6 is that the Internet is in danger of running out of IP addresses. The Internet continues to grow by leaps and bounds and is approaching the 4 billion addresses limitation of IPv4. IPv6 offers a virtually unlimited number of new IP addresses. Here's a sample IPv6 address:

```
EFDC:BA62:7654:3201:EFDC:BA72:
7654:3210
```

IPv6 retains most of IPv4's characteristics, but many important things change for the better. IPv6 provides a wide range of improvements for IP networking, including the following:

- **Big changes in the way you get IP addresses.** In the IPv4 environment, you have to contact your ISP and get a new IP address. After you get the address, you then configure the computer with the IP address information, and update a router. This process can take hours or days depending on your ISP and its technical support. IPv6 uses a new collection of features called autodiscovery, autoconfiguration, and autoregistration. Together, they provide easier management of a network with no manual intervention. By using a system of queries and Plug-and-Play, the network can detect and automatically assign an address. Autoregistration is the way IPv6 handles dynamic adding, which is the process of updating a computer's host name and address information in DNS.

- **Improved security enhancements.** Security services such as packet authentication, integrity, and confidentiality are part of the design of IPv6. Because these security services are built right into IPv6, they are available to all TCP/IP protocols, not just specific ones such as SSL, PPTP, and S-HHTP. This means that your organization can be made more secure easily. The new IPSec protocol, a component of IPv6, adds an additional level of security for IPv6 and creates a secure, TCP/IP-level, point-to-point connection. IPv6 also offers security to applications that currently lack built-in security and adds security to applications that already have security features.

- **Better living through multimedia.** IPv6 provides new capabilities for high-quality, streaming, and multimedia communications, such as real-time audio and video.

Until 1994, class C addresses were the smallest block of IP addresses that could be assigned by InterNIC (Internet Network Information Service). (InterNIC is the organization responsible for domain name registration and IP registration for the Internet.) Many smaller companies that need IP addresses don't need a class C network with 256 IP addresses. Likewise, some companies need more than a class C but less than a class B, and so on. In response to the limitations of A, B, and C classes of IP addresses, InterNIC implemented Classless Internet Domain Routing (CIDR, which is pronounced "cider"). CIDR networks are described as _slash x_ networks, where _x_ represents the number of bits in the IP address range that the InterNIC controls. Table 4-3 lists a sampling of slash x network configurations that support a specific number of IP addresses.

Table 4-3	Common Slash x Network Configurations	
Network Type	*Subnet*	*Number of IP Addresses*
Slash 27	255.255.255.224	32
Slash 28	255.255.255.240	16
Slash 29	255.255.255.248	8
Slash 30	255.255.255.252	4

Subnet subdivision

TCP/IP networks can procreate by subdividing into multiple smaller networks called subnets. A *subnet* is a collection of computers that can communicate with each other without the need for routing. Subnets are created using the host portion of an IP address to create something called a subnet address, or subnet mask. This IP address allows the workstation to identify the network of which it is a part. When you use a DSL router with static IP addressing, you create a subnet. If you use a DSL bridge, your LAN is part of the ISP subnet.

Using private IP addresses

With the proliferation of TCP/IP as the networking protocol of choice for LANs, organizations wanted placeholder IP addresses that they could use for their private networks. The IETF (Internet Engineering Task Force), the group responsible for implementing and maintaining Internet standards, set aside a Class A, B, and C series of IP addresses that can be used exclusively for intranets. These *private IP addresses* can't be assigned on the Internet and will not route through the Internet. Therefore, any organization of any size can use these IP addresses for their intranets. For building your own intranet, this means you don't have to lease IP addresses from an Internet service provider to set up a TCP/IP network.

The address ranges reserved as private IP addresses are as follows:

10.0.0.0 – 10.255.255.255 for Class A networks

172.16.0.0 – 172.31.255.255 for Class B networks

192.168.0.0 – 192.168.255.255 for Class C networks

Using these addresses on your local network makes them invisible to the Internet. These private IP addresses hide behind the single registered IP address used by the router with NAT activated. What's NAT? Read on.

It's all in the translation

Network Address Translation (NAT) is an Internet standard that allows your local network to use private IP addresses, which are not recognized on the Internet. The IP address used for the router as a gateway is provided by the ISP as part of the DSL service. Figure 4-1 shows how NAT works on a DSL connection.

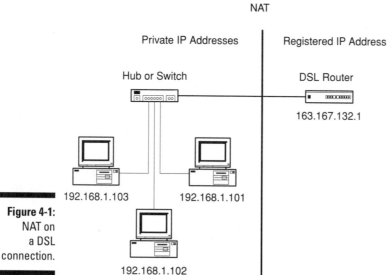

NAT

Private IP Addresses | Registered IP Address

Hub or Switch | DSL Router

| 163.167.132.1

192.168.1.103 | 192.168.1.101

Figure 4-1:
NAT on
a DSL
connection.

192.168.1.102

The computers behind the NAT can access the Internet through the router, but Internet users can't access the computers behind the router. You can allow traffic from the Internet to pass through NAT, however, by mapping ports to specific private IP addresses. For example, you can configure NAT so that it lets Web browser requests from the Internet pass through port 80 to connect to a computer running a Web server on your local network.

Using private IP addresses with NAT, smaller organizations can realize significant cost savings because only a single-user Internet access account is required for connecting an entire LAN to the Internet. NAT also provides increased security because the IP addresses used on the intranet are unrecognizable on the Internet.

NAT works with DHCP to provide dynamic IP addressing. Routers that support both DHCP and NAT let you assign private IP addresses to your internal users from a pool of IP addresses, so you don't have to configure the Microsoft TCP/IP stack on every computer.

Using private IP addresses and NAT can be a cost-effective way to connect a LAN to the Internet, but NAT has some disadvantages. One of the drawbacks of NAT is that you limit direct, two-way, host-to-host communications through the Internet. This means Internet users can't make a direct connection for such applications as video conferencing or voice over IP (VoIP).

Using registered IP addresses

Registered IP addresses are recognized and routable on the Internet. By using registered IP addresses, you can do a lot more with your DSL connection. IP addresses linked to specific hosts and domain names enable Internet users to access a host computer running as a Web server (or any TCP/IP application server) using the user-friendly text identifier, such as www.angell.com. This makes you a provider of Internet services for all Internet users or a private workgroup.

A configuration using a block of registered IP addresses on a LAN might break down as follows:

- ✔ One IP address goes to a Windows NT Server to be used by Microsoft Internet Information Server (IIS) running as a Web server. With a single IP address assigned to the Web server, you can run multiple virtual Web servers off the single IP address.

- ✔ One IP address is assigned to the Windows NT Server to be used by Microsoft Exchange Server for e-mail.

- ✔ Multiple IP addresses are assigned to client computers (hosts) to enable two-way Internet access for IP video conferencing or voice communications and other real-time applications.

You get registered IP addresses from the ISP. The annual fee is anywhere from $25 to $100 for a small block. You also work with the ISP to register your domain name and enter subdomain names into their DNS server. After you get your domain name, you can have the ISP register the subdomains for all your IP addresses on their DNS server.

Using static and dynamic IP addressing

As you've read, you can have private IP addresses and registered IP addresses. Both of these IP address types can be assigned to computers or other network devices on a static or dynamic basis.

A *static IP address* is a fixed IP address assigned to a specific computer or other device on your network. The IP address remains associated with that computer or device so that it can be accessed from the Internet. Think of an IP address as a telephone number that connects to a specific residence. For every computer or device you want available to Internet users, you need an assigned static IP address. For a computer running Microsoft Windows, you enter the IP address as part of configuring Microsoft TCP/IP.

Before you can configure the computer on your LAN for static IP addressing, you need the following information from your ISP:

- ✔ An IP address for each workstation
- ✔ A subnet mask IP address for your network
- ✔ The IP address assigned to your DSL bridge or router, which is the gateway IP address
- ✔ A host and domain name for the registered IP addresses
- ✔ DNS server IP addresses, which are typically IP addresses supplied by the ISP

Using a DSL router that acts as a DHCP server enables dynamic IP addressing — effectively removing the requirement that individual computers must have static IP addresses. The DHCP server automatically manages the assignment of IP addresses, subnet masks, and default gateway addresses to computers as they sign on. The server then manages the IP address table, making sure that only one address is assigned to each active workstation. From a user's perspective, these negotiation and assignment procedures are transparent. You can use DHCP with registered IP addresses or private IP addresses used with the NAT (Network Address Translation) feature incorporated in most DSL routers. DHCP is a typical feature in DSL routers and all Microsoft Windows clients (Windows 95, 98, NT Workstation, and 2000 can act as DHCP clients).

What's In a Name Anyway?

If an IP address is the equivalent of a telephone number on the Internet, a *domain name* is equivalent to the name of the person or organization that the telephone number is assigned to. As people move around, their telephone numbers change but their names do not. Domain names provide the friendly text interface to IP addresses.

The process of converting domain names to machine-readable IP addresses is called *name resolution.* During the resolution process, a computer called a DNS (Domain Name System) server translates the domain name into an IP address.

When you provide an address for most Internet operations, such as pointing your Web browser to a Web site or sending e-mail, you can use either the IP address or the domain name method. Most organizations use domain names as their form of addressing because they are easier to read and understand. Your domain name can be moved to different IP addresses, but the domain name always remains the same.

Host names, domains, and subdomains

Domain names are based on a hierarchical structure. Each period, or dot, within the domain name identifies another sublevel of the overall organization through which the message must pass to arrive at its destination. The order of levels in a domain always proceeds left to right from the most specific to the most general, such as david.sales.angell.com.

No common way exists to decode the meaning of different levels within a name. The person setting up the address defines the nomenclature used in the address.

A *host name* is the name of a specific server on a specific network within the Internet. The host name is the leftmost part of the domain name. For example, www.angell.com indicates the server called www within the network at angell.com. When you use both the host name and the domain name, you're using a fully qualified domain name (FQDN).

Each host belongs to a domain, and each domain is classified further into domain levels. The format of a host name with a domain name is

   HostName.DomainName

The Domain Name System provides a centralized online database for resolving domain names to their corresponding IP addresses.

Domain names can be further subdivided into subdomains. These are arbitrary names assigned by a network administrator to further differentiate a domain name. The format of a host name with both subdomain and domain names is

   HostName.SubdomainName.DomainName

Both the domain and subdomain names further describe a particular computer.

Top-level domain names

At the end of all Internet addresses is a three-letter domain level such as .com or .edu, which is referred to as a top-level domain. These *top-level domains* provide an indication of the organization that owns the address, and they always appear at the end of the domain name. The purpose of the top-level domain is to provide another level of distinction for a full domain address. Seven organizational domains are available, as shown in Table 4-4.

Table 4-4	Organizational Domains
Organizational Domain	*Entity*
.com	For-profit commercial organizations
.edu	Educational institutions
.gov	Nonmilitary government organizations
.int	International (NATO) institutions
.mil	Military installations
.net	Network resources
.org	Nonprofit groups

There are also geographic domains, such as .au for Australia, .uk for United Kingdom, and .jp for Japan. Geographic domains indicate the country in which the name originates. In almost all instances, the geographic domains are based on the two-letter country codes specified by the International Standards Organization (ISO). You'll typically find many domain names that include the .com in front of the two-letter country code, such as www.angell.com.uk.

A new crop of top-level domains

Like so many other areas of the Internet, the pool of available domain names ending with the traditional three-letter, top-level domain name is dwindling rapidly. As a result, a new crop of top-level domain names is available for use on the Internet. The International Ad Hoc Committee (IAHC), which is comprised of Internet standard-setting bodies and legal and communications experts, is implementing seven new top-level domain names.

Table 4-5 describes and lists these new top-level domain names. The IAHC plan establishes as many as 28 competing registration firms to handle the Internet registration of these new domain names.

The domain name game

Your domain name is registered with InterNIC. InterNIC acts as an Internet registration service to keep track of all names and addresses. InterNIC

Table 4-5	The Seven New IAHC Top-Level Domain Names
Top-Level Domain Name	*What It Is*
.arts	Cultural or entertainment organizations
.firm	General category for businesses or other organizations
.info	Entities that provide free informational services
.nom	Individuals' personal Web sites
.rec	Recreational activity sites
.store	Businesses offering goods or services for sale at their Web sites
.web	Organizations that specialize in Web-related activities

handles the registration services for .com, .net, .org, .edu, and .gov top-level domain names.

Names are registered on a first-come, first-serve basis. Registering a domain name implies no legal ownership of the name.

A top-level domain name establishes your business name as defined in the Domain Name System. From the top-level domain, you can register subdomain names directly from your ISP. These are the addresses that define host machines and servers, such as www.angell.com or david.angell.com.

Domain names are not case sensitive, so it doesn't matter whether letters are uppercase or lowercase. No spaces are allowed in a domain name, but you can use the underscore (_) to indicate a space. You can use a combination of the letters *A* through *Z*, the digits 0 through 9, and the hyphen. You can't use the period (.), at sign (@), percent sign (%), or exclamation point (!) as part of your domain name because DNS servers and other network systems use these characters to construct e-mail addresses.

After you have decided on your domain name, you should check to see whether it's available. Because of the immense popularity of the Internet, finding a domain name that isn't already in use is becoming harder. You should do your own checking of domain names before you decide on a domain name and submit it your ISP.

The best way to check the availability of a domain name is to use your Web browser and point it to http://www.rs.internic.net, which takes you to the InterNIC Web site (see Figure 4-2). You can search the domain name database to see whether someone else registered the name you want. If the name is available, the database tells you that no match was found. If the name is registered, the database displays public information about the domain name holder, including the name and address of the holder, contact information, and IP addresses connected with the domain name.

After you know your domain name, the easiest way to get it registered is to delegate the ISP to do it on your behalf. The ISP will register your domain name with InterNIC as part of establishing your Internet service. The ISP may charge a nominal fee to do the registration. InterNIC charges $70 to register the domain name for two years. Registering for a domain name involves providing, through online forms, administrative, technical, and billing contact information as well as IP addresses of the DNS servers handling your domain name. After your domain name is registered, you can make changes to your InterNIC database record through the InterNIC Web site.

When you register your domain name through your ISP, make sure that the registered party is yourself, not your ISP. If you change ISPs, you want to make sure that you own the name and can transfer it to your new ISP.

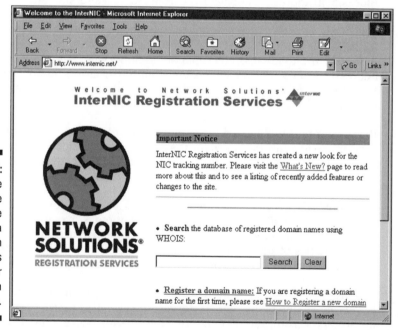

Figure 4-2:
The InterNIC site lets you see whether a domain name is available or has been taken.

IP Configuration Recipes

At this point, you might be totally confused about IP addresses and domain names and what they have to do with your DSL connection. To put this information in a tangible form that you can relate to, here are some real-world IP configuration recipes that you might use when setting up your DSL service. Security issues are also involved in making your local network accessible to the Internet, but for now focus on IP address and domain name issues. Security issues are explained in Chapter 6.

Using a block of registered IP addresses

In the first example, or recipe, you have a block of registered IP addresses and their associated domain names, which means that the gateway device and the computers on the local network are accessible from the Internet.

Blocks of IP addresses are available from most ISPs for a monthly cost based on the number of IP addresses. Some ISPs include a block of IP addresses as part of the service. The IP addresses assigned to you by the ISP are available for use while you are the ISP's DSL customer. They remain the property of the ISP and return to ISP upon termination of the service. When you buy IP addresses from an ISP, the ISP ensures that those addresses are routable on the Internet and should also include custom subdomain names, such as www.angell.com, david.angell.com, and so on.

Bridged, or routed, DSL service requires three IP addresses just for the IP server. One IP address is used for the router, one for the Ethernet connection, and one for the WAN connection. This means that if you get a block of 8 IP addresses, only 5 are available for hosts on your LAN.

For routed DSL service, you can get IP addresses as a block of 8, 16, or 32 IP addresses. These IP addresses typically cost about $25 a month per block of 8 addresses. These blocks of IP addresses are determined by the inherent characteristics of subnet routing. If you're using bridge DSL service, you can get IP addresses in any number because you don't use any subnet routing.

Suppose you have a small network of five computers. One of these computers is a server running Windows NT Server. You also plan to use the computer as a Web server using Microsoft Internet Information Server (IIS). In addition, you want to assign IP addresses to four computers that can be running Windows 95, Windows 98, or Windows NT Workstation as their operating system. As part of your DSL service, you plan to use a DSL LAN modem and use e-mail boxes hosted by your ISP.

One IP address will be going to a Windows NT Server to be used by Internet Information Server running as a Web server. With a single IP address assigned to the Web server, you can run multiple virtual Web servers off the single IP address.

Table 4-6 lists examples of IP addresses and domain names for this scenario. Figure 4-3 shows how this IP address configuration would look on a small network.

Table 4-6	Static IP Address and Domain Name Recipe for a Small Network	
IP Address	**Subdomain**	**What It Is**
209.67.232.2	www.angell.com	The IP assigned to your Web server. When users on the Internet enter the URL www.angell.com in their browser, they connect to your Web server running on the Windows NT Server computer.
209.67.232.3	catie.angell.com	The IP address assigned to a host computer named catie on the LAN using the angell.com domain.
209.67.232.4	david.angell.com	The IP address assigned to a host computer named david on the LAN using the angell.com domain.
209.67.232.5	joanne.angell.com	The IP address assigned to a host computer named joanne on the LAN using the angell.com domain.
209.67.232.6	starr.angell.com	The IP address assigned to a host computer named starr on the LAN using the angell.com domain.

Using NAT and private IP addresses

This section describes a scenario for a small office that is interested in only jacking up the speed of Internet access to a small network of five users. The business is currently using several POTS modem accounts and wants to consolidate them into a single DSL connection. This business will use a DSL router that includes NAT (Network Address Translation) and optional DHCP features. The IP address used for the router as a gateway is provided by the ISP as part of the DSL service.

These machines can access the Internet through the router, but Internet users can't access the computers behind the router. The router, however, can use TCP/IP ports to allow certain types of TCP/IP application traffic, such as port 80 for Web browsers connecting to a Web server on your local network.

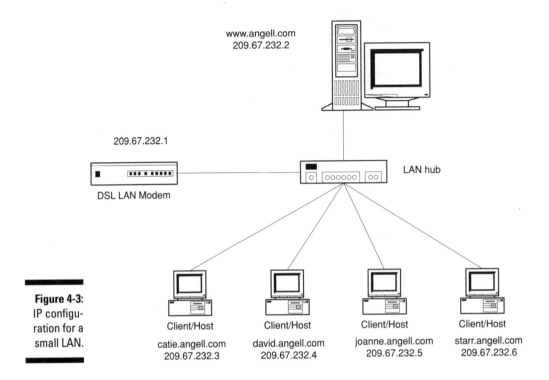

Figure 4-3:
IP configu-
ration for a
small LAN.

www.angell.com
209.67.232.2

209.67.232.1

DSL LAN Modem

LAN hub

Client/Host
catie.angell.com
209.67.232.3

Client/Host
david.angell.com
209.67.232.4

Client/Host
joanne.angell.com
209.67.232.5

Client/Host
starr.angell.com
209.67.232.6

Figure 4-4 shows a LAN-to-DSL connection using a router with NAT activated. Notice that the computers behind the NAT use private IP addresses that aren't recognized on the Internet.

These private IP addresses are not assigned on the Internet and will not route through the Internet. Therefore, using these addresses on your local network makes them invisible to the Internet. These private IP addresses hide behind the single registered IP address used by the router with NAT activated. For example, with five computers on your local network, you can use the private IP address range of 192.168.0.1 through 192.168.0.5 and a subnet mask of 255.255.255.0.

With your private IP addresses defined, you can choose to assign a specific IP address to each computer on your LAN or use the DHCP feature of the DSL router to automatically assign IP addresses from the pool to each computer on an as-needed basis.

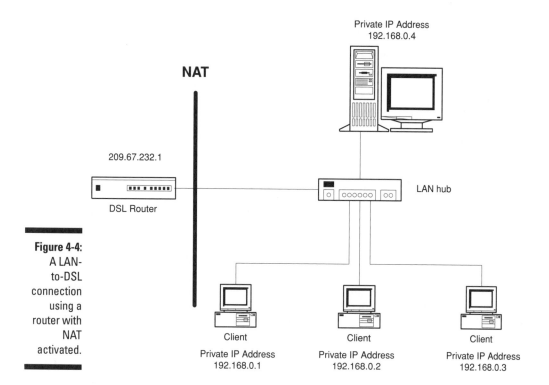

Private IP Address
192.168.0.4

NAT

209.67.232.1

DSL Router

LAN hub

Figure 4-4:
A LAN-
to-DSL
connection
using a
router with
NAT
activated.

Client
Private IP Address
192.168.0.1

Client
Private IP Address
192.168.0.2

Client
Private IP Address
192.168.0.3

PPP on DSL

The Point-to-Point Protocol (PPP) is the TCP/IP protocol that enables dial-up connectivity to the Internet. PPP is the protocol that allows a computer to connect to the Internet to enjoy the benefits of a direct connection and use TCP/IP programs, such as a Web browser and an e-mail client. In Microsoft Windows 95, 98, and NT, the PPP-based connections are handled through Dial-Up Networking (DUN).

Using PPP for DSL connections enables Microsoft Windows users to tap into the same tools they used for analog modem connections to make DSL connections. In the case of using PPP over an always-on DSL connection, each new connection to the Internet is a unique session. The use of PPP by DSL users also helps preserve the large infrastructure built around PPP dial-up networking by ISPs.

Right now there is no standard for using PPP over DSL. Two leading forms of PPP are used for DSL: PPP over ATM and PPP over Ethernet (PPPoE). PPPoE holds the most promise because it dovetails nicely into the existing Ethernet-based networking that dominates PC networking.

PPP over Ethernet

PPP over Ethernet (PPPoE) does what its name implies: It enables dial-up networking capabilities over Ethernet. PPPoE is a software driver that works with an NIC (network interface card) to create a dial-up session through the NIC to the LAN and out through the DSL bridge or router. Because PPPoE uses Ethernet, you can use it to allow multiple PCs to share a single DSL connection.

No standard protocol exists for managing the interaction between a PC (host) and a DSL modem. PPP over Ethernet, however, relies on two widely accepted standards: PPP and Ethernet. PPPoE is an open solution in that it automatically works with existing PC hardware and software, existing Ethernet network interface cards (NICs), and all DSL modems that use MAC (Media Access Control) bridging.

PPPoE was submitted in August 1998 to the Internet Engineering Task Force (IETF) as an Internet Draft. Recently, PPPoE has been submitted as an RFC (Request for Comment). This RFC is a publicly available document that describes the standard proposal and answers questions. Internet Drafts and RFCs are part of the process that enables a technology specification to become an Internet standard.

The significance of PPP over Ethernet is that it makes DSL service installation easier to set up for users and ISPs. ISPs incorporate the PPPoE driver into their setup software. Users install the software with all the DSL connection and ISP information on their computer. After the software is installed, you use Windows Dial-Up Networking to make the DSL-to-Internet connection in the same way you make an analog or ISDN modem connection. From the ISP side, the benefit of using PPPoE is that it allows ISPs to leverage their existing dial-up user infrastructure, such as authentication, usage accounting, and configuration management.

Perhaps one of the most powerful benefits of using PPPoE is that it enables users to access multiple network services from the same DSL connection. This allows ISPs to offer multiple IP services that can be available from the same DSL connection. For example, a user can use an IP connection to a corporate network using VPN over a PPP session during the day. After hours, the user can use another Dial-Up Networking profile to make a connection to the Internet for personal use.

PPP over ATM over DSL

Asynchronous Transfer Mode (ATM) is a high-speed switching technique used to transmit high volumes of voice, data, and video traffic. ATM operates at speeds ranging from 25 Mbps to 622 Mbps. ATM is used mainly in telephone company backbone networks, although large organizations are also using ATM. Adding the Point-to-Point Protocol (PPP) on top of ATM enables TCP/IP traffic to be carried over an ATM network all the way down to a computer without being translated. This configuration requires an ATM adapter card in each computer that connects to a single-user ATM ADSL bridge or a multiuser ATM DSL bridge or router. PPP over ATM isn't a practical solution for most small offices and consumers because ATM adapter cards and ATM DSL CPE are considerably more expensive than Ethernet-based cards and CPE.

Chapter 5

DSL CPE Field Guide

· ·

In This Chapter

▶ Finding out the type of DSL CPE out there

▶ Understanding the fundamentals of DSL CPE

▶ Emerging interoperability

▶ Chipsets, firmware, and DSL

▶ Identifying your single-user DSL CPE options

▶ Pinning down your multiuser DSL CPE options

▶ Changing your single-user DSL connection to a shared one

· ·

*D*SL service ultimately connects your computer or local area network to the Internet or another network. What stands between you and this connection is the Customer Premises Equipment (CPE). The type of DSL CPE you use as part of your DSL service plays an integral role in defining what you can and can't do with your Internet connection.

You need to understand the differences in DSL CPE being offered because DSL is typically sold as a complete package that includes the DSL service, Internet access, and CPE. DSL providers take this approach because they must ensure that the DSL CPE works with their DSLAM (DSL Access Multiplexer) equipment at the CO (central office).

This chapter gives you a solid grounding in your DSL CPE options as they define the capabilities of your DSL connection.

Types of DSL CPE

DSL CPE devices break down into single-user or multiuser solutions. Single-computer DSL CPE consists of the following:

What's a DSL modem anyway?

DSL providers and CPE vendors commonly use the term *DSL modem*. This is more a metaphor than a descriptive term to describe the modem-like functionality of a DSL modem in making the connection to the Internet through DSL.

The danger to consumers in using the term DSL modem is that it masks the different underlying CPE technologies that can be used to connect a single user or multiple users on a LAN. What a CPE vendor or DSL provider calls a DSL modem can be a Universal Serial Bus (USB) external modem, a PCI adapter card, or a DSL bridge.

✔ **PCI (Peripheral Component Interconnect) adapter cards**

✔ **USB (Universal Serial Bus) modems**

✔ **Bridges**

These single-user solutions are usually bundled with dynamic IP Internet access from the ISP. This means that the IP address changes depending on the lease times established by the ISP. Most dynamic IP Internet service doesn't support Domain Name Service, which means that you can't use your own domain name as part of the service. You can get around this, however, by hosting your Web site and e-mail box with a hosting service.

Most of these single-user DSL CPE options can be modified using proxy server software or Ethernet-to-Ethernet routers to enable multiple computers to share the DSL connection. More on these solutions later in this chapter.

DSL CPE for the multiuser network environment consists of the following:

✔ **LAN modems (bridges)**

✔ **Routers**

 Make sure you check out the specifics of the DSL CPE and IP configuration options offered by a DSL provider before committing to the service, especially if you want or plan to use more than a single computer with the service. Many ILEC ADSL and UADSL (G.Lite) offerings use dynamic IP addressing and CPE that restricts access to only a single computer. They do this to control the data traffic moving through your DSL connection.

DSL CPE Interoperability

Interoperability is important in driving down the cost of DSL CPE. Over time, standards will emerge across all DSL flavors, which will enable DSL customers to buy CPE from any vendor. In the current environment, DSL providers control the deployment of CPE to ensure that it works with their DSLAM equipment. To offer consumers more options, a growing number of DSL CPE vendors have interoperability agreements with different DSLAM equipment vendors.

Universal ADSL (UADSL), which is based on the G.Lite standard, will over time enable DSL customers to buy their own CPE in the computer retail channel in the same way that analog modems are sold.

 Different DSL CPE may work with a DSL provider's offerings. If you change your DSL service to another DSL flavor or DSL service provider, however, you might have to shell out money for new DSL CPE and a new round of start-up charges.

Chipsets, Firmware, and DSL CPE

At the heart of any DSL device is its brain, which is commonly referred to as a *chipset.* A DSL device's functionality is determined by what is embedded in its chipset. As standards such as G.Lite become formalized, they're embedded into chipsets to make the DSL CPE comply with the standard.

The trend in hardware is to allow software upgrades to chipsets. *Firmware* is a type of chipset used in computer hardware devices that holds its content without electrical power. Many devices now allow you to download software from the vendor's site to perform a do-it-yourself firmware upgrade. With DSL service and standards still in flux, many DSL CPE vendors include upgrade-able firmware in their devices to allow for changes in the capabilities of their devices.

Single-User DSL CPE Options

Universal Serial Bus (USB) and PCI (Peripheral Component Interconnect) DSL modems are single-computer solutions. Another single-computer DSL CPE is a bridge that uses the Ethernet interface to connect to a computer. This bridge connects to a computer through a network interface card, but unlike its LAN cousins, it supports only a single computer.

DSL CPE based on the G.Lite standard will primarily consist of external USB modems or PCI adapter card modems. The single-computer DSL bridges used today by ILECs will give way to USB modems or PCI adapter card modems.

USB modems and PCI adapter card modems use Microsoft Windows (95, 98, and NT) Dial-Up Networking (DUN) to make connections because they use Point-to-Point Protocol (PPP). This is the protocol that allows a computer to connect to the Internet and use TCP/IP programs, such as a Web browser and an e-mail client. Using PPP for DSL connections enables you to tap into the same tools you use for analog modem connections. In the case of using PPP over an always-on DSL connection, each new connection to the Internet is a unique session.

Perhaps one of the most powerful benefits of using PPP for DSL connections is that it enables you to access multiple network services from the same DSL connection. For example, you can use an IP connection to a corporate network using VPN over a PPP session during the day. After hours, you can use another Dial-Up Networking profile to make a connection to the Internet for personal use. Over time, this type of service may become more widely available.

No standard exists for using PPP over DSL. Currently, the two leading forms of PPP used for DSL are PPP over ATM and PPP over Ethernet (PPPoE). PPPoE holds the most promise because it fits in nicely with the existing Ethernet-based networking that dominates PC networking.

USB: Say goodbye to UART

External DSL modems use a different data communications port than the traditional UART (Universal Asynchronous Receiver/Transceiver) serial port used for POTS and ISDN modems. DSL speeds simply overwhelm the UART's limited 115-Kbps capacity. Universal Serial Bus (USB) is the relatively new data communications port that can support DSL. USB is an open, Plug-and-Play specification for PC peripheral connectivity.

USB ports appeared on PCs in 1996, but few USB devices existed until recently. With the release of Windows 98, USB now has the full support of Windows. Most new machines are equipped with one or two USB ports, and a cornucopia of new USB devices are available. The USB topology allows up to 127 peripherals to be connected to a computer from distances of up to about 15 feet.

USB improves the way you can use peripheral devices, including DSL modems, by providing a better I/O (input/output) port standard than the ones offered by traditional serial, parallel, mouse, and keyboard ports. With USB, installing peripherals is fast and easy: You just plug the cable into the back of your computer. No more messing around with device drivers, add-in cards, or system settings. You don't even have to restart the computer.

In the case of DSL modems, the USB connection is an attractive option because the USB port supports two data rates: 1.5 Mbps and 12 Mbps. At these speeds, USB ports can easily support DSL modem data speeds.

DSL modems that use the USB port are external devices that look much like a standard analog modem, as shown in Figure 5-1.

(Courtesy of Efficient Networks.)

In the PCI modem cards

As with USB connections, DSL PCI adapter card modems are designed as a low-cost, single-computer DSL solution. DSL PCI modem cards are typically less expensive than external USB DSL modems. These cards insert into a PCI (Peripheral Component Interconnect) slot and combine the Ethernet interface to your PC with DSL modem functions.

From your PC's perspective, the DSL modem appears as a network interface card (NIC). DSL NICs plug into PCI desktop computers, eliminating the need for an external modem. The DSL line connects directly to the card. Most of these cards are Plug-and-Play compatible to make installation easy. Figure 5-2 shows the Cisco 605 PCI ADSL modem.

You can create a multiuser connection by installing proxy server software, which makes the DSL modem card act as a proxy server. You need to install another network interface card (NIC), which will connect to the LAN.

Single-computer bridges

A single-computer bridge operates like its multiuser LAN counterpart, except the bridge supports data coming from only a single computer. This type of DSL bridge recognizes data traffic coming from only a specific network adapter card — the one it's connected to. By adding this restriction to a

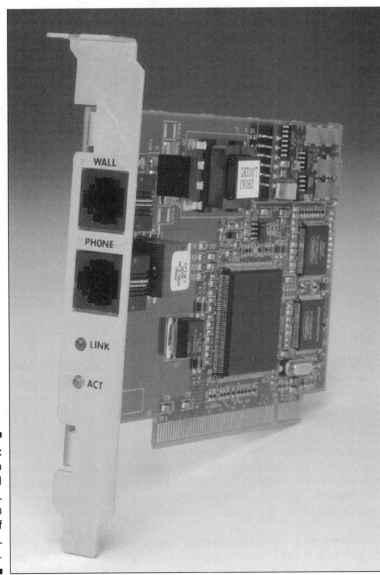

Figure 5-2:
The Cisco
605 PCI
ADSL
modem is an
example of
a PCI DSL
modem.

(Courtesy of Cisco Systems, Inc.)

bridge, the CPE restricts sharing the DSL connection across multiple computers. Many ILECs bundle these single-computer bridges as DSL modems for
their ADSL offerings.

MAC attack: Single-user bridges

MAC, or Media Access Control, is the protocol that operates at the data link layer of the OSI protocol stack. It controls access from the network interface card (NIC) to the network media (cabling). Each NIC has its unique MAC address, which is also called a *hardware address*. (An example MAC address is 00-80-AE-00-00-01.)

Bridges accept TCP/IP packets from NICs. A LAN bridge learns all the MAC addresses of all NICs on the local network to determine what is and isn't local data traffic. A single-computer bridge used for a DSL connection recognizes traffic from only the single MAC address of the NIC it's connected to.

Ways exist for sharing a DSL connection using a single-computer bridge, but don't expect an ILEC to tell you. You can use an Ethernet-to-Ethernet router that sits between your LAN and the bridge or install a software proxy server. More on Ethernet-to-Ethernet routers later in this chapter.

Multiuser DSL CPE Options

The most popular DSL-to-LAN interface is Ethernet. DSL CPE devices that use the Ethernet interface include bridges and routers. These DSL CPE connect to a LAN through a hub or a switch using standard 10Base-T cabling.

DSL bridges are often referred to as LAN DSL modems. Unlike the single-user bridge, a DSL LAN modem connects to a hub, which makes the DSL service available to all the computers connected to the hub or the switch. Any of these network DSL CPE can be connected also directly to a single computer using a 10Base-T crossover cable that connects to a network interface card. Some DSL CPE use a separate Ethernet port for connecting to either a single computer or a hub, so you don't need the crossover cable.

Many LAN modems available today are hybrid devices that incorporate bridge capabilities as well as some router capabilities. Most of these LAN modems, however, don't support as many users as routers and don't support NAT and DHCP.

A DSL router provides more functionality than a bridge, and most DSL routers now include a built-in hub to allow you to build your network and DSL connection in one step.

Bridge over DSL waters

A *bridge* is a simple device that has to decide only whether the packet is intended for the local network or the remote network. A bridge knows the hardware addresses of the NICs of computers connected to the network. The bridge reads the destination hardware address in a packet and decides whether the packet is going to a host on the same network or across the bridge to a different network.

DSL LAN bridges are typically cheaper and easier to install than DSL routers. However, DSL LAN bridges can't provide security for your local network, but routers can. Bridges lack any data-filtering capabilities, so there is no basic checking of incoming packets, which means computers on your LAN have no protection from Internet intruders. A bridge makes your network part of the ISP network, not a separate network, as is the case with a router. However, you can add a proxy server, an Ethernet-to-Ethernet router, or an Internet security appliance to add security to a bridge connection.

Going the DSL router route

Routers work at the Internet, data link, and physical layers of the TCP/IP network model. A router is a more sophisticated gateway device than a bridge. A *router* allows data to be routed to different networks based not on hardware addresses (as in a bridge) but on packet address and protocol information.

This decision-making functionality, called *filtering,* not only enables a router to protect your network from unwanted intrusion but also prevents selected local network traffic from leaving your LAN through the router. This is a powerful feature for managing incoming and outgoing data for your site.

Two inherent features in most DSL routers are Network Address Translation (NAT) and Dynamic Host Configuration Protocol (DHCP). The NAT feature lets you use private (that is, unregistered) IP addresses for your local network workstations. These private IP addresses aren't recognized on the Internet, so computers behind the NAT router remain invisible to Internet users. Using these private IP addresses costs you nothing and also makes your local network inaccessible to outsiders by keeping local traffic separate from Internet data traffic. A DSL router that supports NAT also offers basic security for your local network because NAT blocks incoming access to local computers from the Internet.

DHCP allows the router to assign IP addresses automatically to computers on a LAN as they start up. The DHCP server built into the router dynamically assigns IP addresses from a pool of IP addresses, which can be private IP addresses or public IP addresses (IP addresses recognized on the Internet). For more information on NAT and DHCP, see Chapter 4.

A new generation of DSL routers designed for the small business market provide a cost-effective way to connect a LAN to the Internet. These DSL routers come from such vendors as Netopia, FlowPoint, Cisco, and Cayman. Many of these DSL routers also include a built-in hub, which means you can easily set up a LAN without buying a hub.

DSL routers cost more than DSL LAN modems, but if you factor in the benefits of using a router, DSL routers can be the better deal in the long run.
Figure 5-3 shows the Netopia 7100 router, which is used as CPE for SDSL service from NorthPoint Communications.

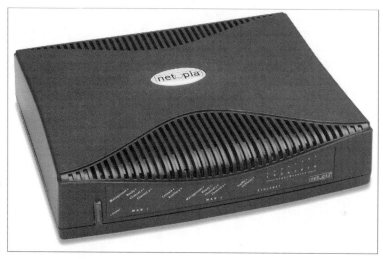

Figure 5-3:
The Netopia
router.

(Courtesy of Netopia, Inc.)

Sharing a DSL Connection

A single-user DSL bridge accepts only Internet-bound data traffic from the
computer with the NIC to which the bridge is connected. You'll find that a
number of ILEC DSL offerings use this single-user CPE solution. The good
news is you can use a device called an Ethernet-to-Ethernet router (also
called a proxy server) to enable you to share the DSL connection using the
single-user bridge.

An Ethernet-to-Ethernet router is an external box that sits between the DSL
bridge and your LAN. It has two Ethernet ports. You connect the 10Base-T
cable coming from your DSL bridge to one port and the 10Base-T cable
coming from your LAN hub to the other Ethernet port. The DSL bridge uses
the Ethernet-to-Ethernet router's MAC (Media Access Control) address as the
single address it recognizes. Behind the Ethernet-to-Ethernet router using
NAT (Network Address Translation) and DHCP are the computers connected
to the LAN.

A number of vendors offer Ethernet-to-Ethernet routers, including Cayman
(www.cayman.com), Netopia (www.netopia.com), and MultiTech Systems
(www.multitech.com). Another important benefit of using an Ethernet-to-
Ethernet router is that it provides some security for a bridge connection.
Bridges don't include any filtering capabilities to screen incoming data traffic
the way routers do.

Chapter 6

To Serve and Protect: Internet Security

*T*he Internet is a double-edged sword. On one side, its open TCP/IP proto-cols and networking environment facilitate information sharing, improve connectivity, and provide greater access to the underlying network. On the other side, the same open protocols and networking systems that make the Internet popular also make security a big issue.

Making an always-on connection to the Internet through DSL demands some security planning to protect your local network from malicious attacks on several fronts. This chapter helps you navigate through the Internet security maze and guides you through a variety of affordable software and hardware solutions.

Internet Security 101

Being connected to the Internet comes with some security risks. In the con-text of DSL service, security is even more important because the Internet connection is always on. The adage "A chain is only as strong as its weakest link" is relevant to security. TCP/IP includes only minimal security features,

You can't be too careful about Internet security. The number of sophisticated hacking tools available on the Internet these days is amazing — and frighten-ing. When your company connects to the Internet, you must decide how to secure your network from intruders.

The mantra of Internet security

The mantra of Internet security is minimizing unsolicited inbound connections. This means that you must control what types of inbound connections you allow.

You need to allow inbound connections for essential Internet services, such as incoming e-mail or Web server responses to Web browser requests from your LAN users. Beyond these basic access requirements, the more inbound connections you allow, the more potential risks.

Internet security options never offer an ironclad solution because security is a constantly evolving game of cat-and-mouse. As hackers develop new ways to break into networks, new safeguards are developed. Adding security to your network is a trade-off between unfettered, two-way access to the Internet and severe restrictions on what you can and can't do with your connection. The other big trade-offs are cost versus risk and performance versus risk.

Connecting your local network to the Internet with TCP/IP can open the doors to a variety of attacks from outsiders to gain access to your private resources or to raise havoc to your local network or Internet connection. The three common forms of TCP/IP-based security attacks follow:

- **Denial of service attack.** The goal of a denial of service attack is not to steal information but to disable a device or a network so that users no longer have access to network resources. Flooding an e-mail box is a common use of this threat. Any type of service for which users rely on timely access is vulnerable to this type of attack.

- **Network packet sniffing.** Programs called *packet sniffers* capture packets from the local area network and display them in a readable manner. The source and destination users of this information probably never even know that this information has been tapped.

- **IP spoofing.** An *IP spoof* is when an attacker assumes the role of a trusted IP address and masquerades as it. In an IP spoof attack, the IP address of a particular machine on the Internet or within an intranet is controlled and managed entirely by the intruder, not by the owner or administrator of that machine.

Enhancing security with IPv6 and IPSec

Security services (such as packet authentication, integrity, and confidentiality) are part of the design of IPv6, which is the next generation IP networking protocol. These capabilities can ensure that packets are from the correct sender, haven't been altered in transit, and cannot be seen by hackers.

Because security services are built into IPv6, they're available to all TCP/IP protocols. The new IPSec (IP Security) protocol is a component of IPv6 and is available for IPv4. The IPSec protocol adds an additional level of security and creates a secure, TCP/IP point-to-point connection. IPv6 offers security to applications that currently lack built-in security and also adds security to applications that already have security features.

Understanding Firewalls and Proxy Servers

The primary purpose of a *firewall* is to control access, which it does by using identification information associated with TCP/IP packets to make a decision about whether to allow or deny access. This decision is based on a set of defined rules that describe which packets or sessions are allowed.

An important difference between various firewalls is the amount and quality of information used to make decisions. The more information that is collected, the less likely it is that an intruder will get through the firewall.

Firewalls differ in their architecture and features. Two primary types of firewall architectures, however, are available: packet filter firewalls and proxy servers. Hybrid firewalls that incorporate a mix of both features are also available.

Packet filter firewalls

A *packet filter firewall* offers basic network access control based on protocol information in the IP packet. This information is compared to a collection of filtering rules when the IP packet arrives at the firewall. These rules specify the conditions under which packets should be passed through or denied access. Most packet filter firewalls are incorporated into routers because packet filtering for restricting access is a logical extension of the router's functionality. As far as Internet users are concerned, the only accessible machine on the inner network is the specified host machine. Packet filter firewalls are generally considered less secure than application-level gateways (proxy servers).

At a minimum, all firewalls use the information found in the IP packet to make decisions, as described in Table 6-1. These basic components are the protocol, destination IP address, destination IP port, source IP address, and source IP port. Some firewalls accept or reject packets by using the destination and source IP addresses and IP port information; other firewalls use the protocol in the packet. The firewall can track packets to determine who may be attempting to access the network and issue alarms to help detect suspicious activity as it is occurring.

Table 6-1	**Components Used by a Firewall in an Internet Environment**
Component	*What It Is*
Protocol	Transmission Control Protocol (TCP) or User Datagram Protocol (UDP)
Destination IP address	Identifies the location of the computer receiving the data transmission
Destination IP port	Identifies the application on the computer that will receive the data transmission
Source IP address	Identifies the location of the computer initiating the data transmission
Source IP port	Identifies the application on the computer initiating the data transmission

Proxy servers

Proxy servers, which are also referred to as *application*, or *session-aware*, *firewalls*, go beyond the basic packet-filtering mechanism. They accept or reject data traffic based on an entire set of IP packets that are part of an entire session to the same address. For example, Web browsers like Internet Explorer will generate multiple data packets as they request files from various parts of the Internet. These individual data packets are all part of the larger session. Session-aware proxy servers provide better security than just plain packet-filtering firewalls.

Session-aware firewalls shuttle information from the original connection to the second connection. They sit between your LAN and the DSL bridge to the

Internet. In essence, the proxy server masquerades as the destination computer to the network client and as the network client to the destination computer.

Proxy servers provide these key benefits:

- ✔ **Sharing your DSL connection.** A proxy server allows multiple computers on a LAN to share Internet access from a single IP address. This means that you can share a DSL connection offered as a single-user solution for your entire LAN.

- ✔ **Securing your LAN from outside intruders.** A proxy server provides tight security. Attacks based on IP spoofing can't reach the local network. Computers on the LAN access the Internet indirectly through the proxy server.

- ✔ **Better utilization of your DSL connection.** Most proxy servers also include Web caching. A Web-caching proxy server cruises the Web and examines pages that your LAN users have visited and that have been cached on the server. If a page has been modified, the proxy server stores a new version on a local drive. It can also use certain guidelines to hit links on that page to pull down related pages. This prevents your users from having to access the Internet for frequently used resources, saving time.

Microsoft Proxy Server

Microsoft Proxy Server 2.0 is a full-featured firewall product that delivers controlled Internet access and monitoring of Web usage. As a BackOffice family member, it's designed specifically for Windows NT Server. Microsoft Proxy Server 2.0 lists for $995.

Microsoft Proxy Server also includes a Web cache server to improve the performance of your DSL connection by cutting down on the number of requests the client needs to generate for servers on the Internet. Microsoft Proxy Server includes support for NAT (Network Address Translation).

Proxy Server 2.0 delivers firewall-class security through its packet-filtering capability. Besides being resistant to common attacks, such as IP spoofing, Proxy Server provides packet filtering and access control to block users behind the proxy from accessing certain sites and resources. This feature lets an administrator reject specific packet types at the IP level before they reach higher-level application-layer services.

Enabling packet filtering causes Proxy Server to drop all packets sent to a destination, except those that match a list of predefined packet filters. You create a filter for the packet types you want the Proxy Server to accept. Microsoft has defined a set of reasonable default packet filters.

Proxy Server can also alert you to suspicious activity at the packet level. For example, if the proxy server rejects more than 20 packets in one second (the default), you're alerted that your network may be under attack.

Proxy Server includes a new feature called *dynamic packet filtering.* As you might recall, a packet filter does its job based on a TCP service and a port number. For example, a Web server typically uses port 80. A packet filter must always be listening at port 80 for any traffic bound for the Web server. Therefore, the system always has an open port that an intruder could exploit. In dynamic packet filtering, the proxy listens to port 80, but the port is not truly open. When a request is made at port 80 for an HTML document, for example, the proxy opens the port to allow the packet through. As soon as the conversation is over, the proxy closes the port, and the system is locked down again.

WinProxy

For smaller organizations, WinProxy (`www.winproxy.com`) is a good choice. WinProxy is one of the easiest to install and configure, and it costs a fraction of the price of Microsoft Proxy Server. WinProxy sells for between $99 and $299, depending on the number LAN users, and runs on Windows 95, 98, or NT. WinProxy can operate with either a fixed or a dynamic IP address on the Internet gateway (bridge or router). WinProxy handles most Internet protocols, including POP3, FTP, HTTP, DNS, NTTP, and IMAP4.

WinProxy's HTTP proxy service can support incoming as well as outgoing HTTP requests, allowing you to use WinProxy as an incoming firewall to a Web server located on your LAN. WinProxy lets you restrict Internet access to specific client PCs by IP address, and you can fine-tune Internet access by restricting users to specific protocols. WinProxy's URL filtering feature enables you to block an unlimited number of Internet destinations. This blacklist of taboo Internet sites goes into a text file that the server reads on bootup.

WinGate

WinGate 3.0 (`www.wingate.com`) is another affordable proxy server that runs on Windows 95, 98, and NT. It's flexible, and many shortcomings found in previous versions have been dramatically improved. Improved installation wizards simplify setup and greatly reduce configuration requirements. WinGate 3.0 also includes default security configurations so that users can get their system up and running quickly. The WinGate Gatekeeper provides a unified interface to managing your proxy server.

Security Solutions for the Rest of Us

Traditionally, Internet security products were designed and priced for the needs of the enterprise computing market. These firewall products were

targeted at larger organizations because they had high-speed Internet connections and needed the security to protect their networks. Most of these
firewall products are simply too complicated and too expensive for smaller
organizations.

With the advent of affordable high-speed connections through DSL, a growing
number of companies are now offering firewall solutions targeted at smaller
organizations. A variety of hardware and software security products are available. These security solutions break down into the following categories:

- **DSL routers (hardware).** Most DSL routers come with built-in filtering
  capabilities as well as NAT (Network Address Translation) to add basic
  firewall functionality. These devices are built with routing in mind, however, so their security options are limited. The NAT feature of DSL
  routers, which offers firewall protection to a LAN by using private IP
  addresses, also enables sharing a single-user DSL connection across the
  local network.

- **Proxy servers (software).** The software proxy server runs on a computer with two NIC cards installed. One NIC sends traffic to the DSL CPE,
  and the other NIC connects to the local network. The proxy server
  works as the gateway between traffic from the local network and traffic
  going to the outside network. As with DSL routers and hardware proxy
  servers, using a proxy server lets you share the DSL connection if you're
  using a single-user bridge.

- **Proxy servers (hardware).** These devices, which are also called
  Ethernet-to-Ethernet routers, sit between a DSL bridge and your LAN. A
  bridge is essentially a dumb device that moves all TCP/IP traffic from
  your LAN to the Internet and vice versa without any filtering. The proxy
  server box includes two Ethernet ports. One connects to the DSL bridge,
  and the other connects to the LAN via the hub. The proxy server works
  as the gateway between traffic from the local network and traffic going
  to the outside network. Adding an Ethernet-to-Ethernet router to a DSL
  bridge connection not only provides security functions but also allows
  you to share a single-user DSL connection across a LAN.

- **Firewall appliance (hardware)**. These are firewall boxes that provide
  the highest level of security by combining sophisticated packet filtering
  and proxy server services as well as by providing basic routing functionality. A new generation of affordable firewall appliances targeted at
  smaller businesses offers sophisticated firewall capabilities in a box.

- **Protocol isolation.** This solution is the most restrictive and requires
  each computer to have two NIC cards. Using the two NIC cards lets you
  bind a different protocol to each NIC. For example, one NIC uses
  Microsoft's NetBEUI networking protocol for the LAN, and the other NIC
  uses TCP/IP to make the Internet connection.

In this section, I describe each of these security options in more detail.

DSL routers as firewalls

DSL routers offer basic firewall protection because they use filtering to route TCP/IP traffic. The firewall is built by establishing filters within the DSL router. A *filter* is a rule you establish on the router to train the router to drop or let pass certain packets. Nothing less intelligent than a router at a minimum can do this because of the software required. The three fields that can trigger the filters are the source address, the destination address, and the port number. Dedicated firewall products go beyond the basic filtering capabilities of routers.

The source address answers the question, "Where did this packet come from?" The destination address answers the question, "Where is this packet going?" When the destination is one of your computers, you can decide whether any packet is allowed to go there.

Sometimes you may choose to redirect the packet to a special server. In this case, the filter says "The packet is okay, but send it to a specific server, such as an e-mail or Web server." A port number answers the question, "What server is this packet destined for?" If you're not running a particular server, why should you accept packets destined for a nonexistent server? For example, if you're not running a Web server, you can freely and safely ignore any HTTP packets trying to connect to a Web server. Your filter says, "I refuse to talk about that subject."

NAT as a firewall

Network Address Translation (NAT) is a common DSL router feature that enables you to create a basic firewall. NAT is an Internet standard that allows your local network to use unregistered IP addresses, which are not recognized on the Internet, and also make the Internet connection with an IP address that is recognized on the Internet. The NAT feature creates a temporary connection between the private IP address and the Internet-routable IP address. Because NAT uses private IP addresses for the local LAN, computers beyond the DSL router using NAT are invisible to the Internet.

NAT not only provides added firewall security for your LAN but also results in significant cost savings because only a single-user Internet access account is required for connecting an entire LAN to the Internet.

The downside of using NAT with private IP addresses is that it restricts the use of your Internet connection. Because you aren't using registered IP addresses, the computers on your LAN can't be directly linked to the Internet for activities (such as running servers) or applications (such as video conferencing). You can, however, map incoming data traffic to a server operating behind NAT (such as a Web server) by using port mapping. Port mapping is explained in the next section.

Application control through port mapping

You control TCP/IP (that is, Internet) traffic by defining what ports your DSL router allows to pass through. As you recall from Chapter 4, a *port* is an address that identifies the application associated with the data. Ports are typically transparent to network users because many ports are universally assigned to specific TCP/IP services. Port numbers below 255 are defined as defaults for a variety of universal TCP/IP applications and services. For example, the universally recognized port for HTTP (the Web) is port 80, and FTP uses port number 21. Port mapping incoming data passes through the NAT router and onto the specified computer on the local network.

Using port mapping in the DSL router, you can redirect one type of service, such as HTTP (Web), to a single host on the local network behind the NAT firewall. This mapping allows the router to redirect any packets from a Web browser via the Internet to a Web server running on a computer on the LAN. The Web user on the Internet uses the IP address (and its domain name, if any is assigned to the IP address) as the URL for the Web server on the local network.

Most DSL routers are capable of passing TCP connections on specific ports, but others offer only limited capabilities. Port mapping features become important if you plan to use such TCP/IP applications as video conferencing or IP voice. These applications typically use a unique combination of dynamically assigned ports. If the DSL router isn't capable of passing through TCP traffic over dynamically assigned ports, you won't be able to use a video conferencing program like Microsoft NetMeeting or CU-SeeMe for audio or video conferencing. Make sure that your DSL router has the capability to support dynamic port pass-through.

Software proxy servers

Proxy software makes a computer with two NICs installed the proxy server for the LAN. One NIC is connected to the LAN hub, and the other is connected to the DSL bridge. A proxy's job is to accept requests from a machine on the internal network, screen it for acceptability according to specified rules, and then forward it to a remote host on the Internet.

The leading software proxy servers are Microsoft Proxy Server 2.0, which runs on Windows NT, and WinGate Pro and WinProxy, both of which run on Windows 95, 98, and NT. WinGate Pro and WinProxy are less powerful but also less expensive.

Proxy server in a box

A growing number of vendors are offering proxy servers in a box. These devices are commonly called Ethernet-to-Ethernet routers and work in a similar manner to their software cousins. An Ethernet-to-Ethernet router sits between your DSL bridge and your LAN to manage TCP/IP traffic. These boxes don't provide the LAN-to-WAN connection; you need to use the DSL bridge that is part of your DSL service to make the connection to the Internet.

An Ethernet-to-Ethernet router includes two Ethernet ports, which are used to separate the network physically into two areas. The WAN port attaches to the DSL router or bridge, and the LAN port attaches to the network hub or switch. These devices add the power of routing to your LAN-to-DSL connection. They provide a combination of filtering along with application-level proxy server capabilities. And like software proxy servers, they allow you to share a single-user DSL bridge connection across an entire LAN. Most Ethernet-to-Ethernet routers also support NAT and DHCP.

Ethernet-to-Ethernet routers, like most DSL routers, enable you to redirect various Internet services, such as HTTP (Web), to a single host on the local network behind the NAT firewall. This feature allows the router to redirect any packets from a Web browser via the Internet to a Web server running on a computer on the LAN. The Web user on the Internet uses the IP address (and its domain name, if any is assigned to the IP address) as the URL for the Web server on the local network.

And also like DSL routers, Ethernet-to-Ethernet routers support port mapping, but some offer only limited capabilities. Port mapping features are important if you plan to use TCP/IP applications, such as video conferencing or IP voice. These applications typically use a unique combination of dynamically assigned ports. If the DSL router isn't capable of passing through TCP traffic over dynamically assigned ports, you won't be able to use a video conferencing program (such as Microsoft NetMeeting or CU-SeeMe) for audio or video conferencing. Make sure that the Ethernet-to-Ethernet router you're considering includes the capability to support dynamic port pass-through. More on working with Ethernet-to-Ethernet routers in Chapter 5.

A growing number of vendors are selling Ethernet-to-Ethernet routers, including MultiTech Systems' Proxy Server, Netopia's 9100 Ethernet-to-Ethernet Router, and Cayman System's 2E Series.

Security as an appliance

An *Internet appliance* is a hybrid device that incorporates both packet filtering and proxy server capabilities in one device. A growing number of companies offer security appliances, especially for Virtual Private Networking (VPN). A few of these companies are Redcreek (www.redcreek.com), VPNet (www.vpnet.com), and Sonic Systems (www.sonicsys.com).

Sonic's approach to their products is unique. SonicWALL products build from a sophisticated, yet easy-to-use, firewall foundation and then add basic IP networking features, such as DHCP and NAT. SonicWALL devices are about the size of an external modem and have two Ethernet ports on the back panel. You connect your LAN to one of the ports and the DSL bridge to the other port.

Sonic Systems offers a line of SonicWALL products including SonicWALL, SonicWALL Plus, SonicWALL DMZ, and SonicWALL VPN. The SonicWALL and SonicWALL Plus products provide firewall and routing capabilities. SonicWALL DMZ includes a DMZ port to allow Internet users to access public servers, such as Web and FTP servers. The DMZ feature allows you to use registered IP addresses for Internet-recognized servers for specific computers on your LAN while protecting other computers on the LAN from Internet attacks. SonicWALL VPN offers an all-in-one VPN, firewall, and content filtering box. Using a SonicWALL VPN box at each end of a DSL-to-Internet connection creates a secure private network between any two remote sites.

Following are key features of all SonicWALL products:

- ✔ **Firewall security.** SonicWALL uses sophisticated packet inspection technology found in enterprise-level computing. SonicWALL protects against denial of service attacks, such as the ping of death and IP spoofing.

- ✔ **Hacker attack prevention.** SonicWALL is automatically configured to detect and thwart denial of service attacks, such as the ping of death, SYN flood, LAND attack, and IP spoofing.

- ✔ **Java, ActiveX, cookie, and proxy blocking.** Although Java and ActiveX offer advantages, such as enhancements to Web pages, they can also be used by hackers to steal or damage data. SonicWALL can examine HTTP traffic and block the Java and ActiveX portions of a Web page download as well as block cookies. You can customize what trusted sites users can download Java and ActiveX components from. When a proxy server is located on the WAN, it's possible for LAN users pointing at this proxy server to circumvent content filtering. SonicWALL can disable access to proxy servers located on the WAN.

✔ **Internet content filtering.** This feature allows you to create and enforce Internet access policies tailored to the needs of your organization. The feature uses a Content Filter List that is updated automatically on a weekly basis. The SonicWALL includes configurable content filtering. This allows you to block incoming content. You can also make certain users exempt from filtering. When a user tries to access a forbidden site, you can deny access, create a log, or both. You can even create personalized warning messages for each user on your LAN.

✔ **IP address management.** Network Address Translation (NAT) translates IP addresses used on a private LAN to a single registered IP address, which is used for the Internet access. This adds a level of security because the address of a PC on the LAN is hidden from the Internet. This product also includes built-in DHCP server and client capabilities. SonicWALL supports one-to-one NAT, which allows you to have registered IP addresses operating behind the NAT firewall. The SonicWALL creates a passage through the wall to the specific computer using the IP address.

✔ **Web browser management.** You can configure and monitor the SonicWALL by using a Web browser.

✔ **Network access rules.** This feature allows you to extend the SonicWALL's firewall functions. For example, a rule can be created that blocks all traffic of a certain type, such as Internet Chat (IRC), from the LAN to the Internet. Another rule could be created to give Internet users access to a server on the LAN, such as a public Web server.

✔ **ICSA certified.** ICSA provides an ICSA Firewall Certification, which is a kind of Good Housekeeping seal of approval.

✔ **Upgradeable.** SonicWALL uses upgradeable firmware to allow for constant updates; it even notifies you when a new upgrade is available.

✔ **Optional IPSec VPN.** SonicWALL VPN Upgrade provides an easy, affordable, and secure way for a business to connect offices or teleworkers. Using data encryption and the Internet, SonicWALL VPN provides secure communications between two or more sites.

✔ **Web Proxy Relay.** This feature allows SonicWALL to redirect all Web requests transparently to the proxy without client configuration.

✔ **Monitoring and alerts.** SonicWALL maintains a log of events that may be security concerns. You can be alerted when an attack to an e-mail account or an e-mail pager is in progress.

✔ **Intranet support**. This feature enables you to restrict access to certain resources on the LAN.

Isolating your protocols

A simple approach to securing your local network is isolating your local network from Internet traffic by using a different networking protocol. This approach requires the use of two NICs in every PC. One NIC directs TCP/IP packets to a router or a bridge that communicates with the Internet, and the other NIC directs packets to the network running a different networking protocol, such as the Microsoft NetBEUI networking protocol for Windows.

By setting up two cards, each bound to a different networking protocol (TCP/IP and NetBEUI), you can ensure that the packets from one network can't mix with packets from the other network. In this way, the two networks operate separately on the PC.

The downside of this approach is that the two networks are isolated from each other, so two network interface cards must be installed and configured in each computer.

Part II
Shopping for DSL

The 5th Wave   By Rich Tennant

"Troubleshooting's a little tricky here. The route table to our destination hosts include a Morse code key, several walkie-talkies, and a guy with nine messenger pigeons."

In this part . . .

ADSL connection opens up the world of high-speed Internet access to the bandwidth deprived. This part gives you a game plan for navigating through the DSL shopping maze to get to the pipe dream.

You get help in evaluating TCP/IP applications that take advantage of high-speed, always-on DSL. You check out DSL ILEC and CLEC offerings and ISP partners. Wrapping up, you get the lowdown on what IP services — such as IP addresses, domain name services, Web hosting, and e-mail hosting — to consider as part of your Internet connection.

Chapter 7

Getting Your DSL Connection

· ·

In This Chapter

▶ Creating your guide through the DSL service maze

▶ Checking out DSL availability in your area

▶ Calculating your need for speed

▶ Shopping tips for DSL-based Internet service

▶ Figuring out what DSL will cost

· ·

*A*DSL connection opens up the world of high-speed Internet access to the bandwidth deprived. Getting to that world, however, can be tricky business in the chaotic Wild West atmosphere of today's DSL service deployment, packaging, and pricing. Before you go shopping for DSL service, read this chapter to get a framework for navigating through the DSL shopping maze.

Creating Your DSL Game Plan

The consumer adage *caveat emptor* (buyer beware) is alive and well when it comes to shopping for DSL service. Hype abounds, and just below the surface lies a host of fine-print traps waiting for the unsuspecting. What you don't know can cost you a bunch of time and money. Your first order of business is to create a DSL game plan that will guide you through the process of selecting your DSL connection to the Internet.

The big three pieces of DSL service

Three key parts make up a typical DSL service offering, although most DSL service is sold as a complete Internet access and connectivity package. Each of these main pieces of DSL service breaks down into sublevels of related issues. The three top-level components of the total DSL ecosystem follow:

- **The DSL circuit provider.** An ILEC or a CLEC delivers the DSL pipeline service. They operate the DSLAMs (Digital Subscriber Line Access Multiplexers) at central offices and link them to a backbone from which the ISPs then connect. Different DSL circuit providers offer different DSL flavors and speed capabilities.

- **Internet service.** The Internet service provider (ISP) is often your interface to DSL service. The ISP handles the TCP/IP networking infrastructure associated with your DSL connection. ISPs buy DSL circuit services from CLECs and ILECs. What you can and can't do is defined in large part by what services the ISP offers.

- **Customer Premises Equipment (CPE).** The specific CPE you use with your DSL service plays a pivotal role in defining what is and isn't possible with your Internet service. The DSL circuit provider determines which products can be used because the CPE must work with the DSLAM equipment used by the DSL circuit provider.

DSL game plan checklist

The process of establishing a DSL connection involves evaluating interrelated components that make up your DSL service package. Here is a checklist of key questions you need to address as part of your DSL shopping game plan:

- **Is DSL service available in your area?** This is the essential first step. If DSL service is available in your area, you need to assess what DSL flavors are being offered. In areas with competition for DSL service, you may have a choice of different types of DSL technologies, such as SDSL and ADSL. Each of these technologies has different capabilities, pricing, and configuration options.

- **What are your bandwidth needs?** Articulating your bandwidth needs is, at best, an educated guess, but you'll need a consistent benchmark for comparative shopping. Determining your bandwidth needs depends on a variety of factors. In the real world, the biggest constraint on bandwidth is cost. The more bandwidth you want, the more it will cost. One-time start-up and CPE costs typically remain the same (or rise only marginally) as the capacity of the DSL connection increases, so the real cost increase is in the monthly DSL circuit charge. One of the nice features with many DSL flavors is the capability to upgrade your bandwidth without buying new CPE. Bandwidth is addictive, and over time, you'll want more.

✔ **What type of CPE do you need to use?** No DSL service shopping list is complete without an analysis of your CPE options. Typically, the DSL provider drives what DSL CPE can be used with the DSLAM at the CO. The biggest divide in DSL CPE is whether it supports only a single user or multiple users on a LAN. As part of your CPE decision, you also need to think about how you plan to configure your office for IP networking.

✔ **How good is the Internet service provider?** The ISP is in the driver's seat in defining your Internet service capabilities through DSL. The ISP providing the DSL service can put restrictions on your DSL service in its contract. For example, many ISPs set limits on the total amount of data traffic that can pass through your connection in a given month based on the speed of your connection. For example, a 160-Kbps connection might be allowed 10 gigabytes a month, but a 416-Kbps connection might be allowed up to 30 gigabytes. Some ISPs restrict you from running any kind of server on your DSL connection. Because the ISP is also your interface to TCP/IP networking elements, you'll want to know about an ISP's technical support. Does the ISP offer technical support 24 hours a day, 7 days a week?

✔ **What kind of TCP/IP configuration do you plan to use?** You need to think about your TCP/IP networking infrastructure and how it will relate to the DSL-based Internet connection. For example, using registered IP addresses allows Internet users to access servers running on your LAN by using DNS (Domain Name System) addresses, such as `www.angell.com`. Likewise, you'll need an IP address for each host or network device that you want to be accessible from the Internet. For example, if you plan to use video conferencing, you'll want an IP address assigned to the computer running the desktop video conferencing system.

✔ **How many users do you want to connect to the DSL connection?** The number of users (computers) with which you want to share the DSL connection affects the type of DSL ISP service you want. Many ILECs package DSL Internet access service offerings that don't support multiple computers or DNS services. These solutions can be modified to support multiple users (with proxy servers or Ethernet-to-Ethernet routers), but you won't be able to provide any Internet services, such as running a Web server or video conferencing.

✔ **What will the complete DSL service package cost?** You need to crunch all the numbers to get an accurate estimate of what the DSL connection is really going to cost. The total price you pay for your DSL service depends on many variables, including one-time installation charges, CPE costs, and monthly service charges.

Checking DSL Availability

DSL's deployment is on a CO-by-CO basis, which results in a patchwork of DSL service availability. If you live in a large metropolitan area with high concentrations of potential DSL customers, chances are good that an ILEC, a CLEC, or an ISP acting as a DSL CLEC is offering DSL service in your area. If you live in a smaller city, a town, or a rural area, your chances of having access to DSL service are much smaller.

DSL service deployment is measured by the number of COs with DSLAMs up and running. When a CO is activated for DSL service, all customers serviced by that CO have access to the service, which is referred to as *customers passed.* According to TeleChoice, a telecommunications consulting firm, as of the end of 1998, ILECs have installed DSL service in 788 COs that pass a potential 19 million customers. CLECs have installed DSL service in 434 COs that pass a potential 11 million customers. ISPs acting as CLECs have installed DSL service in 12 COs that pass 1 million customers.

The big CLECs have targeted for DSL deployment the larger metropolitan areas, including Atlanta; Boston; Chicago; Dallas; Denver; Detroit; Houston; Los Angeles; Miami; New York; Philadelphia; San Francisco; San Diego; Seattle; and Washington, D.C. DSL service deployment records for ILECs vary considerably from one ILEC to another. Most ILECs, however, are planning large-scale DSL deployments, mostly in the form ADSL or UADSL service.

Skimming the cream

The relationship between CLECs and ILECs is tenuous. ILECs are not big fans of the Telecommunications Act, which enabled the CLECs to enter their telecommunications turf. In ILEC circles, they frequently complain that CLECs are skimming the cream of the telecommunications business by going after the large metropolitan markets and focusing on the business market. ILECs and their supporters negatively refer to CLECs as *cream skimmers.*

From the DSL consumer standpoint, the emergence of CLECs is a powerful force for ensuring that DSL service is aggressively deployed and competitively priced. In many areas, CLECs have already rolled out DSL service, while ILECs have nothing to offer. For example, at the time this book was written, Bell Atlantic (an ILEC) wasn't offering DSL service in major metropolitan markets until mid-1999. On the other hand, the three leading CLECs — NorthPoint Communications, Rhythms, and Covad Communications — are more aggressive in offering DSL service.

Tracking down DSL service

The Web is the number one resource for checking the availability of DSL service, and the best starting points are ILEC and CLEC Web sites. Your local ILEC will typically have, at its Web site, information about its DSL service offerings. Some of these sites include Web forms for checking out DSL availability based on your area code and exchange (prefix) number. Table 7-1 provides a list of ILEC and CLEC Web sites for starting your search for DSL service in your local area. (See Appendix A for a comprehensive list of DSL service providers, including ILECs, CLECs, and ISPs.)

Table 7-1	ILEC and CLEC Web Sites
DSL Provider	*Information Source*
Ameritech (ILEC)	`www.ameritech.com/products/data/adsl/`
Bell Atlantic (ILEC)	`www.bell-atl.com/adsl/`
Bell South (ILEC)	`www.bellsouth.net/external/adsl/`
Covad Communications (CLEC)	`www.covad.com`
NorthPoint Communications (CLEC)	`www.northpoint.com`
Pacific Bell (ILEC)	`www.pacbell.com/products/business/fastrak/dsl/`
Rhythms Netconnections (CLEC)	`www.rhythms.net`
Southwestern Bell (ILEC)	`www.swbell.com`

CLECs typically offer DSL services through partnerships with selected Internet service providers. CLECs provide links to these ISPs at their site, which you can then use to check out their DSL service offerings. Figure 7-1 shows the NorthPoint Communications service areas page, and Figure 7-2 shows the Covad Communications service availability page.

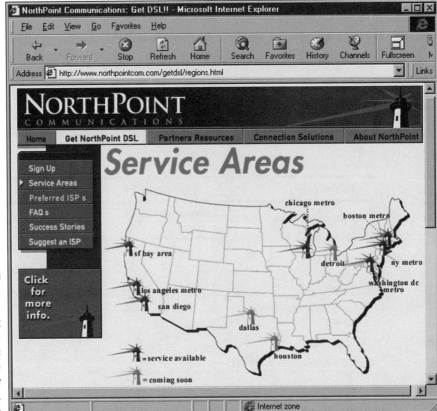

Figure 7-1:
The
NorthPoint
Communi-
cations
service
availability
map.

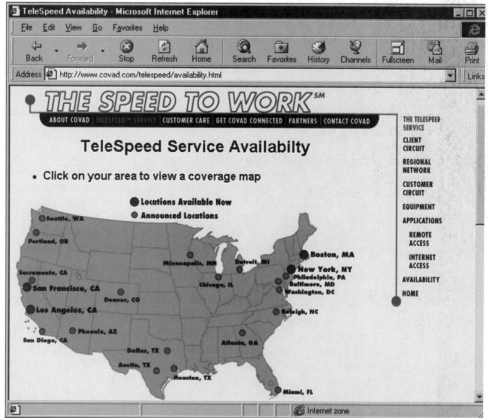

Figure 7-2:
The Covad Communications service availability map.

Several sites on the Internet can help you in your search for DSL service. Internet.com's "The List" provides the most comprehensive database of ISPs at `http://thelist.internet.com`. You can search by area code or other queries. If an ISP offers DSL, the list tells you and provides a link to the ISP's Web site. The `www.xdsl.com` site, which is operated by TeleChoice, a telecommunications consulting company, offers information on the latest DSL happenings.

Have your phone number and address ready

DSL availability is determined with two key pieces of information, your telephone number and the location of your premises relative to the CO. After you determine that DSL service is available in your metropolitan area, you need to get more specific as to whether DSL service is available from the CO servicing your area.

The first check for DSL service to your location is determined by using an existing telephone number already in place at the premises. Specifically, the area code and exchange prefix (the three digits after the area code) hone in on the CO servicing your area. The DSL provider matches this information against the deployment of their DSLAMs. If the DSL provider has a DSLAM installed at the CO servicing your area, you pass the first hurdle. All this tells you, however, is that a DSLAM is installed at the CO.

The next piece of the availability puzzle is determining the distance of your premises from the CO, where the DSLAM is located. This is where the address of your premises comes into play. As you recall, DSL service is distance sensitive. The farther away your premises are from the CO, the fewer speed options are available. If your location is relatively close to the CO, which is the case in many urban areas, you'll have access to the full range of DSL speeds being offered. If your premises are located at a distance beyond the range of some of the higher speeds, you'll have fewer speed options. If your premises are located beyond the distance threshold for the DSL service at any speed, you're out of luck.

A CO is not the only place for DSL capability

Multiple dwelling units (MDUs) is telecommunications jargon for a high-rise apartment or office building. Many of these larger buildings are signing deals to bring fiber into the building and put a DSLAM in the basement to carry high bandwidth up to the tenants in the building.

In these MDUs, you get high bandwidth, even though DSL is nowhere else in the area. As much as 15 percent of the office and high-rise residential space in New York City is already under contract for this arrangement.

Although using the combination of your telephone number and address provides a reasonable guess in determining whether you will have access to DSL service, a variety of technical reasons might, in the end, mean that DSL won't be in your future. The quality of the copper wiring, interference in the ILEC's cable plant, and other technical issues can put an end to your DSL dreams. Ultimately, the only way to know for certain is to have a technician come out to the site and attach test equipment to the circuit, which is part of the DSL service installation process.

Behind the scenes in DSL local loops

The process of verifying the quality of a local loop for supporting DSL service is referred to as *loop qualification*. Setting up DSL service always involves the ILEC, even if the DSL service is being provided by a CLEC. The ILEC switches the copper pair from its system to a DSLAM at the CO. The ILEC is responsible also for bringing the telephone line to the network interface device (NID) at your premises (if it's a new line).

DSL services from CLECs usually involve a new line installed at your premises. Many ILECs offering ADSL or UDSL service can convert an existing line to a high-speed and POTS line. If you order a new DSL line from the ILEC, the ILEC may or may not require you to get POTS service with the line also, which adds to the cost. If you're using a CLEC as your DSL provider, the CLEC must wait until the ILEC does the necessary loop wiring. The ILEC can take from a few business days to a few weeks to set up the local loop for a CLEC. For example, Bell Atlantic takes five days, but Pacific Bell takes ten days.

After the wiring has been brought to the NID, the next phase of the process begins. The installer from the CLEC comes to your premises to do the inside wiring required to bring the line from the NID to the location of the DSL CPE. The CLEC installer tests the line and then connects the DSL CPE to the DSL line. The DSL CPE is typically configured by the ISP and sent to your premises before the inside wiring is performed.

Figuring Out Your Need for Speed

Bandwidth capacity planning is one of the most important and most difficult tasks you'll undertake in setting up a DSL connection. A number of factors come into play when trying to evaluate your bandwidth needs. Here are some basic bandwidth questions you need to answer:

✓ **Will you be running any servers on your DSL connections?** A big factor in determining your bandwidth requirements is whether you plan to run any kind of servers on your DSL connection. If you plan to run a Web server, an e-mail server, or any kind of Internet server, you need to take into consideration the incoming traffic as well as your outgoing traffic. Chapter 8 goes into more detail on calculating bandwidth needs for running a Web server.

✓ **Do you plan to use IP voice and video conferencing?** If used heavily, these cool (but bandwidth hungry) applications can eat up your DSL connection. Supporting multiple simultaneous users compounds the demands.

✓ **How do you plan to use your DSL connection?** If you are going to do symmetrical applications (such as running servers), the upstream of the ADSL connection needs to meet the minimum qualifications. So if you will have IP video and want a 384 Kbps/384 Kbps connection, a 1.5 Mbps/384 Kbps ADSL connection is still fine. Where you run into problems is when your symmetrical application requirements exceed the upstream of ADSL. Most CLECs offer SDSL, which is symmetrical service.

✓ **Do you plan to connect one computer or a network to your DSL line?** If you want to connect more than one computer, you're looking at a LAN-based connection to the Internet. The DSL CPE used in your connection combined with your IP addressing options define what you can do with your DSL connection. You'll find DSL offerings, especially by ILECs, that offer attractive DSL speeds but use single-computer CPE with dynamic IP addressing to restrict the use of the service. You can, however, use a proxy server or an Ethernet-to-Ethernet router to share the connection.

✓ **Can you upgrade to a higher speed?** One of the nice features of DSL service is its bandwidth scalability. This means you can upgrade to a higher speed over time without having to start all over again. For example, SDSL service offers a range of bandwidth options that range from 160 Kbps to 2.3 Mbps. For example, you can start with 384 Kbps and move up in increments to 2.3 Mbps. Most SDSL bridges and routers allow you to upgrade with only minor software configuration changes. DSL service providers will charge a fee for upgrading, typically in the range of $100 to $200.

Some DSL CPE bundled with DSL service packages offered by ILECs are bridges that connect only one computer to the Internet through DSL. Third-party proxy server and Ethernet-to-Ethernet router solutions let you get around this restriction to leverage your DSL connection by sharing it. See Chapter 5 for more information.

Table 7-2 provides some guidelines for bandwidth requirements for different types of Internet applications. You can use these as a rough measure of what different uses demand in terms of bandwidth ranges. Multiply the bandwidth requirements for each computer you want to connect.

Table 7-2	Bandwidth Requirements for Different Applications
Application	*Minimum Bandwidth Requirement*
Desktop video conferencing	384 Kbps
Distance learning	384 Kbps
Interactive video games	128 Kbps
Web surfing	128 Kbps
IP voice	56 Kbps
Teleworking	144 Kbps
Video on demand	1.5 Mbps
Web hosting	384 Kbps – 6 Mbps

DSL and Quality of Service

Bandwidth by itself is not the only determining factor in choosing a connection. It's also important to take a look at an Internet service provider's performance. If an ISP's network is being slowed down by large amounts of packet loss or long delivery times, called *latency,* you are not getting the delivery rate you're paying for. The service provider's own network can create a bottleneck that can hamper your connection, no matter what the bandwidth between the ISP and your premises.

Some CLECs and ISPs offer Quality of Service (QoS) guarantees for their DSL service. What these usually boil down to, however, is a guarantee only on the bandwidth from the DSLAM to your CPE.

Checking Out Your DSL Service Options

Depending on where you are located and the way DSL service is delivered in your area, the Internet service provider is typically the single interface to the DSL service, the CPE, and the Internet access account. Behind the ISP are CLECs and ILECs providing the circuit service and defining what CPE is used for the service. Chapter 10 goes into more detail on the IP networking considerations as part of your Internet service package.

You typically receive a single bill from the ISP for both the Internet service and the DSL circuit. This is the way most CLEC DSL services are sold. This is the easiest approach and the one that makes the most sense. In the case of ILECs, they may or may not offer ISP service as part of their DSL offerings, or they may provide the DSL connection in partnership with ISPs. Some ILECs bill the customer directly for the DSL line, separately from the Internet service provider.

Finding a good Web site

One of the most important sources of information in shopping for DSL service is the Internet service provider's Web site. A good Web site can save you a great deal of time. Unfortunately, the quantity and quality of information provided on DSL services at an ISP's Web site vary widely. Chapter 9 provides a detailed listing of ILEC and CLEC Web sites as a starting point for checking out DSL service. Here are some guidelines to judge a helpful Web site:

- **Does the Web site provide specifics on the ISP's DSL service offerings?** Glossy marketing copy doesn't make for good consumer information. Information packaged to educate the consumer about the service and product details is helpful.

- **Does the Web site include the costs of the ISP's DSL offerings?** Installing and using DSL service involves a variety of charges. A Web site should provide a breakdown of the costs for getting the DSL service, including a menu of optional services.

- **Does the Web site include the ISP's terms of the service?** Unfortunately, most ISPs don't post their terms and conditions on their Web sites. These documents are the fine print of your DSL service. Terms and conditions are usually provided as part of the formal quote.

- **Does the Web site provide good CPE information?** CPE is at the heart of your DSL service capabilities and defines your TCP/IP application options. The Web site should provide coverage of the different CPE options and their prices.

✔ **Does the Web site provide information on IP addresses services and Domain Name Service?** A good Web site will list your IP address options and costs, such as the additional costs for IP addresses and DNS registration. The site should also include a menu of custom services, such as ISP support for running your own e-mail server.

Getting the quote

Shopping for DSL means picking up the telephone and calling an ISP to get a quote for the DSL service you want. Your experience in dealing with an ISP can range from the good to bad. Many ISP staffs know their stuff and do a good job, but others don't. You want to get a formal quote that provides the following information:

✔ **The DSL service and speed you specified**

✔ **The CPE used for the service and its cost**

✔ **The breakdown of one-time set-up charges**

✔ **Total monthly recurring charges**

✔ **Additional charges for custom and IP address services, Domain Name Service, and any additional e-mail boxes**

✔ **A copy of the terms and conditions of your DSL service**

A terms and conditions document spells out what you can and can't do with your DSL service. These contracts are very important, and you need to read them carefully.

The devil is in the details

The terms and conditions in the DSL service contract can often make or break a DSL service deal. Restrictions on your DSL service are spelled out in this contract, so you must read it carefully to understand fully what you can and can't do with your DSL service. Here are some important things to look for in a DSL service contract:

✔ **What is the time commitment for your DSL service?** Many ISPs require that you make a commitment to the DSL service of one to three years. Depending on the provider, the cost per month may drop if you commit for a longer period. Consider whether you think prices will go down in the future before committing to a longer-term contract. DSL service is new, and prices will probably come down as competition heats up. You may not want to pay today's rate for DSL for the next three years. Check also for any early termination charges.

✔ **Is usage restricted?** Many ISPs include restrictions on the amount of data going through your DSL pipe per month per line. For example, there might be a 10-gigabyte limit for a lower-bandwidth DSL connection. Any data moving across your DSL line in excess of the limit costs extra, usually based on a price per megabyte. You may find an even harsher restriction that forbids you from running any servers or using a router that supports NAT and DHCP to share the connection.

✔ **What are the payment terms?** Increasingly, ISPs are billing to credit cards to cut down on their accounts receivable overhead. Others will bill you on a monthly or annual basis, with the annual billing method offering a discount. Many ISPs also require a deposit to start the service.

✔ **Are there any quality of service guarantees?** Most ISPs don't offer Quality of Service (QoS) guarantees for DSL service at this point. DSL is usually sold as a best effort service. However, most CLECs are offering SLAs (Service Level Agreements) for the connection between the CPE and DSLAM. As DSL matures, you'll probably see a premium business class form of DSL service, in which the ISP guarantees a certain level of service for a higher price.

Crunching the Numbers

Shopping for DSL comes down to answering the question, What will it cost? The total price you pay for your DSL service depends on a variety of factors, but the main costs associated with DSL service are the one-time start-up and CPE charges and the recurring monthly charges. Figure 7-3 provides a worksheet you can use to calculate the complete cost of your DSL service.

DSL Service _____

One-time start-up costs

      ISP activation   _____

      DSL circuit activation   _____

      Onsite installation   _____

      CPE purchase   _____

      Other fees   _____

Total One-Time Start-Up Cost   _____

Monthly costs

      ISP service   _____

      DSL circuit service   _____

      CPE lease   _____

      Other fees   _____
        (IP address services, e-mail boxes, etc.)

Total Monthly Cost   _____

Figure 7-3:
DSL service
cost work-
sheet.

Start-up charges

Start-up costs are where you'll take the biggest one-time hit for DSL service. The fixed charges for setting up DSL service don't vary much from one speed to another within the same DSL flavor. The biggest variable cost will be in the monthly service charges for different speeds. Following are the charges you can expect for starting DSL service:

- ✔ **DSL circuit activation and on-site installation.** These charges include the activation of the DSL line to your premises. These costs typically range from $125 to $375 or more, depending on the speed of your service. On-site installation typically includes testing the circuit and the inside wiring and connecting the DSL CPE to the DSL line. LAN integration and setup is not included; you can do this or use the ISP's engineering support services, which can be expensive (they typically bill at $125 per hour).

- ✔ **Internet access account activation.** Setting up your ISP connection also typically has one-time charges for setting up the accounts you need for e-mail, IP address services, Domain Name Service, and any custom services. Beyond a minimum number of IP addresses provided with the DSL service, there will be an additional charge for the first year's rental of more IP addresses. You may have to pay a registration fee if the ISP registers your domain name with InterNIC, which is a separate charge from the InterNIC fee. You may also be charged additional one-time costs for setting up e-mail accounts, Web hosting services, and any other special service you request.

- ✔ **DSL CPE.** The cost of DSL CPE depends on the type of CPE you use and the ISP's markup. Single-user CPE is less expensive than multiuser DSL CPE (bridges and routers). DSL CPE typically costs from $350 to more than $1000. Some ISPs offer modem leasing or rental as part of the connection. Typical DSL routers should be under $995, and typical DSL LAN modems should go for under $495. Single-user USB modems should be under $299, and PCI adapter modems should be under $199. Some ISP may try to charge a router configuration fee.

Monthly recurring charges

Monthly fees are usually a single flat rate charge that is based on the speed of your DSL connection. The two main components of your monthly bill for DSL service are the circuit charge from the ILEC or CLEC and the Internet service cost. The circuit charge is typically part of a single bill from the ISP or, in some cases, on a separate bill from the circuit provider. The Internet service part of your monthly charge may have a base flat rate charge and a usage charge based on megabytes for data traffic exceeding the limit. You may also have other monthly charges for such services as e-mail boxes or Web hosting services. If you're renting your CPE, you'll also have a monthly rental fee.

Chapter 8

TCP/IP Applications for Your DSL Service

*T*he combination of high speed and an always-on DSL-to-Internet connection opens up an intoxicating number of TCP/IP application possibilities. But before you rush off to load applications onto your DSL connection, you need to evaluate your options and their consequences. This chapter takes you through your TCP/IP application possibilities with an eye to their effect on your DSL connection, your time, and your money.

Stop, Take a Deep Breath, and Think

The temptation to try out all kinds of cool, sophisticated client and server applications on your DSL connection is compelling. Stop, take a deep breath, and think. The rule of thumb about TCP/IP applications is: Just because you *can* do it doesn't mean you *should* do it. Each application includes pros and cons that need to be looked at carefully.

Here are some guidelines to keep in mind as you explore your TCP/IP application options and formulate your strategy:

✔ Take a thoughtful strategy that compares the benefits to the commitment it takes to fully utilize an application. There are always trade-offs between an application's benefits and costs.

✔ Be careful if you plan to run any kind of server (such as a Web server or an e-mail server) on your DSL connection. Running a public Web server on your LAN, for example, can make big demands on your DSL connection as well as on your time to set up and administer the server.

✔ IP video conferencing is a sexy application that holds a lot of promise, but it's bandwidth hungry and can eat up your DSL connection.

✔ Expand your DSL connection speed. A cool thing about DSL is that you can upgrade your speed to support growing demands without changing your DSL CPE or your IP address infrastructure. This means your services can evolve as you move up the TCP/IP application learning curve.

✔ Consider using a hosting service for your Web server, e-mail, and other Internet services. Hosting a Web site and e-mail boxes with your domain name is affordable, and many packages are available.

Harnessing your DSL connection

A fatter pipeline to the Internet is only part of the DSL connection story. The high-speed and always-on characteristics of DSL service change your Internet experience and open up new capabilities, including the following:

✔ **Faster Web surfing.** This core function of DSL service makes the Web come alive, and the always-on feature of DSL means instant access. Faster Web surfing means you can do more in less time, which can make you more productive. It also opens up all kinds of new multimedia choices for users in the form of streaming video, graphics, and so on.

✔ **Rapid downloads and uploads.** The more reliable digital nature of DSL service combined with raw speed makes the movement of programs and files more feasible. Downloading a 72MB file takes 25 minutes at 56 Kbps, 10 minutes at 128 Kbps, and 48 seconds at 1.5 Mbps. Your DSL connection becomes your gateway to getting and managing software. Beware, though, because with this new downloading power come new virus risks.

✔ **Running a Web server.** DSL can support running your own Internet Web server on your LAN. You can use the Web server also as the center of your intranet. Running your own Web server enables you to do more things to your site than you can with a Web hosting service, but it demands more management and development skills. You can run a public or a private collaboration Web server.

✓ **Running an e-mail server.** Running an e-mail server means that you become the post office for your domain name on the Internet. The server handles bulk mail coming from the Internet (or on the intranet) and then sorts it to the correct e-mail boxes. With an e-mail server, you can operate a variety of e-mail services, such as using an autoresponder for queries or automatic forwarding of e-mail.

✓ **IP voice and video communications capabilities.** Until DSL, IP voice and video conferencing simply wasn't a viable communications tool. The high speed and always-on features of the DSL connection combined with routable IP addresses means that you can become accessible to people connected to the Internet. An IP address becomes the equivalent of your telephone number, which means you can receive as well as make calls.

✓ **Virtual private networking (VPN).** In a virtual private network, you connect to another computer over the Internet and use encryption to protect the data transmissions going on between the two computers. VPN enables teleworkers to link up to a company network through the Internet and work securely, as if they were sitting at a LAN client at the office.

Monitoring Your DSL Performance

The bandwidth you're buying to access the Internet through DSL is the speed supported by the DSL line as it passes through different points. These points are from your premises to the DSLAM at the CO, and from there to the DSL provider's (CLEC or ILEC) backbone. ISPs make money from subscribing customers, so there's always the temptation to *oversubscribe,* that is, to sign on more customers than you have the bandwidth to support.

A big issue when signing up for DSL service is the backbone capacity of the ISP between their network and the Internet. The connection between the ISP to the DSL provider's backbone and between the ISP to the Internet may be too small to support the volume of traffic created by their DSL users. ISPs can face a situation where the low cost of DSL connections to users doesn't even cover the cost of supporting them on their backbones.

Many ISPs use the sales pitch that their backbone is bigger than that of their competitors. Unfortunately, no independent survey of the size of different ISP backbones exists. After you get your DSL-to-Internet service, however, you can use certain tools to monitor the throughput of your DSL connection. (*Throughput* is an overall measurement of a data connection in terms of real-world speed.) A variety of factors affect the throughput of a DSL connection, including distance, switching components (such as routers), network traffic congestion, electrical interference, and server speeds.

This section takes a look at some popular tools for monitoring the performance of your DSL and Internet connection.

Packet loss, latency, and throughput

Any data communications link experiences bandits of bandwidth. These bandits are the result of a variety of factors that affect the true throughput of a connection. *Throughput* is an overall measurement of a data connection in terms of real-world speed. Factors affecting the throughput of a DSL connection are distance, switching components (such as routers), network traffic congestion, electrical interference, and server speeds.

Latency and packet loss can have a dramatic effect on your Internet connection. *Packet loss* happens when a packet doesn't make it to its destination within a reasonable amount of time and therefore must be resent. *Latency* is the time it takes to get data from one point to another. Packet loss and latency can negatively affect the throughput measurement. An ISP's latency is how long it takes data to travel across its network to the Internet.

Using ping and tracert commands

Microsoft Windows 95, 98, and NT include the ping and the tracert (trace route) commands. These commands let you measure the latency of your IP connections and check the status of an IP address.

The ping command is like a sonar command; it sends a signal across the network and measures the response from the pinged host. The tracert command is an extension to ping in that it lets you trace the route that your data is taking to get to a specific site. It shows you how many routers (called *hops*) are between your computer and the server you specified. If you're using Windows 95 or 98, check out Chapter 12 for an explanation of the ping and tracert commands. If you're using Windows NT, see Chapter 13 instead.

Visual Route

Visual Route is a Windows-based tracer route program that automatically analyzes connectivity problems and displays the results of a trace route on a world map. Visual Route processes all IP hops in parallel instead of consecutively, which produces dramatically faster results. Using Visual Route, you can

- ✓ Perform round-trip and remote trace routes
- ✓ Determine whether a connectivity problem is due to your ISP, the Internet, or the host you are connecting to
- ✓ Fight SPAM e-mail by using contact information provided by pop-up network and domain WHOIS information
- ✓ Identify a problem with a particular network
- ✓ Detect routing loops

Most of Visual Route works without proxy support provided that your firewall does not block ICMP (Internet Control Message Protocol) packets. ICMP is a TCP/IP protocol used to report network errors and to determine whether a host is available on the network. The ping utility uses ICMP.

You can check out Visual Route at the following address:

```
www.visualroute.com
```

Net.Medic

Net.Medic is another tool for monitoring your Internet connection. It provides a visual presentation of vital signs, with reports on throughput, retrieval time, and Web server load and efficiency, as well as network delays and congestion levels. For example, using Net.Medic, you can identify the source of a network bottleneck, such as your ISP, the Internet backbone, or the remote Web server.

Net.Medic can automatically correct some performance problems. For example, Net.Medic can detect a hung server condition before you know it exists. When this happens, Net.Medic can reinitiate your server request, eliminating long waiting times, manual refreshes (retries), and time-outs. Net.Medic uses a problem-fixing technology called AutoCure. After identifying a problem and describing it, Net.Medic will ask whether you want the problem automatically fixed.

The Net.Medic window, shown in Figure 8-1, provides a visual monitoring tool that can do the following:

- View and print a variety of reports on your Internet performance
- Track data transfer in real time at a glance
- Index each Web page's retrieval time to determine whether delays are caused by the Internet, the Web server, or any other Internet server
- Monitor and optimize the online performance of your PC
- Track traffic levels, delays, and the overall performance of your ISP
- Isolate network problems to the Internet backbone
- Analyze real-time Web site performance, including the Web server's response times

Figure 8-1:
The
Net.Medic
window.

For more on Net.Medic, check out the following site:

```
www.vitalsigns.com
```

Running a Web Server

A DSL connection combined with routable IP addresses and DNS means that you can run an in-house Web server. However, you should seriously evaluate the pros and cons of running your own Web server versus using a Web hosting service. A Web site demands bandwidth, and if your site becomes popular, it can easily overload your DSL connection.

You also need to address a variety of administration and security issues in running a Web server on your LAN. If you decide you want to run your own Web server, you'll need Web server software and a powerful computer that can handle the demands of many simultaneous users. You'll also need to do some bandwidth calculations before you order your DSL service.

Going with a Web hosting service

Web site hosting is a commodity business, which means you can get some pretty good deals. The hosting service takes care of the bandwidth and server foundations; you create and manage the content. Today's Web hosting services offer a full range of services to enhance your Web site's capabilities, such as adding database support and enabling e-commerce.

Many DSL ISPs offer low-cost Web hosting. You can also shop for third-party Web hosting services. Two leading Web hosting services are Interland and Highway Technologies. Their Web sites follow:

```
www.interland.net
www.highway.com
```

To find other Web hosting services, check the Web or the back of just about any computer magazine.

Even if you don't use your domain name as part of your DSL Internet service package, you can use a domain name for the Web hosting site, and in most cases you can add e-mail hosting services as well. You can easily manage a Web site remotely using a number of Web site management and content authoring tools, such as Adobe's PageMill, Macromedia's DreamWeaver, or Microsoft's FrontPage.

Web server bandwidth requirements

Running a Web site changes everything in your bandwidth calculations. You can do calculations based on file sizes, the number of hits, or the number of users to get an idea of the required bandwidth capacity for running a Web server. This section provides some formulas to help you define the bandwidth requirements for running an in-house Web server.

Calculations based on file size

The first step in calculating the bandwidth requirements for your Web server is to estimate the average file size of your Web pages. An average Web text page is about 500 words, or about 7K, but as soon as you add a graphic or two, the size increases rapidly. A number somewhere between 30K and 50K is a reasonable starting point. You can use this number if you have not yet designed any of your Web pages.

You can estimate the average file size of your Web site also by multiplying 8 bits for each character in a document. A full page of text has 80 characters per line times 66 lines, which equals 42,240 bits per page. For every 8 bits of data transferred, add 4 bits of overhead. Multiply 42,240 bits per page times 1.5 bits used to transfer 1 bit of data, and you get 63,360 bits per page transferred. This calculation assumes a solid text page with no graphics. Also add the file sizes of any graphics or other elements to your estimate to get a better estimation.

To calculate bandwidth based on a given average file size, you convert the number of bytes in your default-sized file into bits by multiplying the number of bytes by 12 (8 bits in a byte plus 4 control bits), and then express the results in terms of seconds. If your data is recorded in bytes per hour, multiply the number of hours by 3,600, which is the number of seconds in one hour. For example, if your plans call for 500,000,000 bytes to be transferred over a 12-hour period, the equation becomes

500,000,000 bits x 12 / (12 hours x 3600 seconds)

which becomes

6,000,000,000 bits / (43,200 seconds x 1024 bits)

which finally calculates a bandwidth of

135.6 Kbps

Calculations based on the number of hits

Another important calculation for estimating your bandwidth capacity needs is to estimate the number of times a day you think your site will be visited, or *hit*. This is difficult to estimate, but you can make some sensible assumptions. You will probably see a large number of hits as people discover your new site. Then, as time goes on and with continued promotion, you will see an increase. At some point, the hit rate will level off. To get an idea of the traffic that all this involves, you can use the hit rate you expect along with the average size of your Web pages with this formula:

(Expected number of hits per day / Number of seconds in 24 hours) x (Document size in K x 12)

which leads to the following formula:

(Expected number of hits per day / 86, 400) x (Document size in K x 12).

For example, if you expect your Web server to attract 2,000 connections in an average day, and your default page size is 70K, the formula becomes:

(2000 / 86,400) x (70 x 1024 x 12)

which generates the following result:

19.9 Kbps

Calculations based on the number of simultaneous users

To calculate the number of simultaneous users that a given data communications capacity can support, you can do the following calculation. Assuming that you want to stay within the 5-second transmission time for a page of text, and assuming a text file size of 63,360 bits transmitted for the user to receive the page, divide the 63,360 bits by 5 seconds = 12,672 bits per second per user. Then, divide the connection speed by the bits per second per user.

For instance, for a DSL connection of 1.54 Mbps:

> 1,540,000 bps / 12,672 bps per user = 121 simultaneous users on the connection

Web server software

Two classes of Web server software are available: those that run on Windows NT Server and those that run on Windows 95 and 98. Microsoft's Internet Information Server (IIS) runs on Windows NT Server. Microsoft's Personal Web Server (PWS) runs on Windows 95, 98, and NT Workstation. If you want to run a serious Web server on Windows 95, 98, or NT Workstation, you should consider a more robust Web server program such as WebSite Pro.

Giving away the industrial-strength IIS Web server is the Microsoft way to crush its Web server competitors, mainly Netscape. Netscape offers a suite of Web server products, including the Netscape FastTrak Server for smaller organizations and the Netscape Enterprise Server for large-scale Web sites. You can check out these products at the following address:

```
www.netscape.com/server/
```

Apache HTTP Server is the most popular Web server on the Internet because of its UNIX roots, and it's available also for Microsoft Windows 95, 98, and NT. The Apache HTTP Server, which is available for free, is a technically demanding product.

Microsoft Internet Information Server 4.0

Microsoft Internet Information Server (IIS) 4.0 runs on Windows NT Server and offers an impressive collection of features. The price for IIS is also impressive: It's free. Internet Information Server (IIS) 4.0 is included with Windows NT Server or is available as part of the Windows NT Server 4.0 Option Pack, which you can download from the Microsoft Web site:

```
www.microsoft.com
```

IIS 4.0 includes not only Web and FTP server software but also a cornucopia of support and development tools. With IIS 4.0, you can create an unlimited number of virtual hosts on a single IP address or create multiple Web sites attached to different IP addresses. You can use Microsoft FrontPage 98 or 2000 with IIS to create Web site content and manage your IIS Web sites. IIS supports FrontPage Server Extensions so that you and others can update a Web site's content directly through your TCP/IP network or over the Internet.

IIS also uses Active Server Pages (ASP), which allows you to create dynamic content using scripts embedded on Web pages that can be processed on the server or client side. ASP provides an easier-to-use alternative to CGI

(Common Gateway Interface) so that content developers can combine HTML, scripts, and reusable ActiveX server components into Web pages. ActiveX was developed by Microsoft and was primarily designed for developing interactive content for the Web.

Microsoft Personal Web Server (PWS)

Microsoft defines the Microsoft Personal Web Server (PWS) as a desktop Web server. It runs on Windows 95, 98, and NT Workstation and is designed for supporting a small intranet server or as a development staging platform before uploading your site content to a Web hosting service. As is the case with IIS, PWS is free for the downloading or comes bundled with many Microsoft products, such as FrontPage, Internet Explorer, and Windows 98.

PWS includes a number of features that make setting up a Web site at home or in small offices relatively painless. The PWS includes the Personal Web Manager window, which lets you manage the Web server software, several Web site traffic monitoring tools, a guest book and drop box to allow you to interact with Web site visitors, and full support for Active Server Pages.

PWS creates a directory of files available to others but prevents access to other directories on your computer. You can publish multiple directories, including virtual directories, as well as define permissions for each directory, such as allowing files to be read-only. You can add a simple form-based two-way communications system at the Web site. And of course, PWS works with FrontPage 98 and FrontPage 2000. You can even create interactive forms that allow PWS to interface to a database, such as Microsoft Access 97 or 2000.

O'Reilly's Web Site Professional

O'Reilly Software's WebSite Professional is a reasonable solution for running a full-featured Web server on a Windows 95 or 98 computer, but it will cost you $799. You can try it before you buy it by downloading a fully functional 30-day version. Check out their site at the following address:

```
website.oreilly.com
```

WebSite Professional is designed for small businesses and includes the Web server and several other components that provide virtually everything you need to develop and run a Web site. It allows you to run multiple virtual Web servers and supports Microsoft's Active Server Pages (ASP). WebSite Professional also includes support for adding HTML front ends to SQL and ODBC databases, as well as tools for monitoring Web site structure and traffic.

Running an E-mail Server

E-mail service is an essential component of any Internet connection. The benefits of running your own e-mail server include saving on the monthly charges for e-mail box hosting and being able to offer more sophisticated e-mail services to your LAN users. The downsides of running your own e-mail server are that it can be expensive and technically demanding.

Installing and running your own e-mail mail server on your network allows you to manage your Internet e-mail along with your LAN e-mail. If you elect to run your own mail server, mail for your domain will be sent directly from the Internet to the mail server on your network. Connections are made directly from the mail sender on the Internet to your mail server; servers at the ISP are never touched. Typical configurations include providing a mailbox for each user locally on the mail server, or implementing mail-forwarding and redirection of a user's mail to a separate internal or external mail server where the user's mailbox actually resides.

Microsoft Exchange

Microsoft Exchange is a client/server messaging system that is part of the BackOffice suite of Windows NT Server applications. Microsoft Exchange lets you build a powerful messaging system for e-mail on your LAN as well as for Internet e-mail traffic. Microsoft Exchange is pricey ($799 for five clients) for small businesses, however, and requires sophisticated administration.

Exchange lets you set up an e-mail server for your LAN as well as for Internet e-mail. You can also create forms for data input to a database, set up a Microsoft Calendar server for Microsoft Outlook clients, route faxes to the PSTN or through the Internet using Net faxing, and provide a network-news type of threaded discussion, real-time chatting, and a platform for third-party voice mail programs.

Deerfield's MDaemon

MDaemon is an SMTP/POP3 e-mail server software package for Windows 95/98 or Windows NT and is designed for smaller organizations. MDaemon starts at $99 for five mailboxes and goes up to $489 for unlimited mailboxes. Check them out at the following address:

```
www.mdaemon
```

MDaemon provides its e-mail server functionality by using aliasing. All e-mail with the domain name of a company, such as angell.com, is sent to a single POP3 e-mail box. MDaemon can receive e-mail messages sent to this singular

POP3 account and distribute them to the organization's staff. For example, suppose an organization has the domain `angell.com` and an accompanying e-mail box `mail@angell.com` hosted at their ISP. The organization arranges for their ISP to *alias,* or forward, all e-mail messages sent to the `angell.com` domain — regardless of username — to the `mail@angell.com` POP3 account. So all e-mail messages received by an ISP with the `angell.com` domain name, such as `david@angell.com`, `joanne@angell.com`, and `starr@angell.com` are deposited in the `mail@angell.com` mailbox on the ISP's server. MDaemon logs on to this single POP3 account, sorts the messages by username, and then sends them to the matching POP3 mailboxes you've defined in MDaemon. The users log on to their e-mail accounts using their e-mail clients as they would if the e-mail boxes were on the ISP mail server.

MDaemon can send and receive e-mail to the ISP at any user-defined interval as low as a minute. MDaemon works with a DSL connection of any speed and flavor. MDaemon includes a host of other e-mail server features including mailing lists, autoresponders, autoforwarding, multiple domains, and remote administration. MDaemon works with any Winsock-compliant POP3 e-mail client software.

Eudora WorldMail Server

The Eudora WordMail Server is an e-mail server targeted at small to medium-sized businesses. It can be used as both an intranet and Internet mail server. WorldMail includes an SMTP mail and POP3 server. WorldMail Server will work with any Internet standards-compliant e-mail clients but is especially fine-tuned for Eudora Pro and Eudora Lite, Eudora's mail client programs. For more information, check out the following site:

```
www.eudora.com
```

Show and Tell through IP Video Conferencing

IP video conferencing adds exciting show-and-tell capabilities to Internet communications to enhance brainstorming and collaboration. Video conferencing for face-to-face meetings over DSL can save money and time by reducing travel. It can be used for helping teleworkers stay in the loop at the office, for remote expert support, customer service, recruitment, distance learning and training, and telemedicine. The downside of video conferencing is that it demands a lot of bandwidth to get an acceptable quality. A face-to-face meeting via IP video conferencing begins at a 384 Kbps symmetrical connection.

Video conferencing is a mature TCP/IP application in terms of well-established standards, thus ensuring interoperability between different vendor's products. Desktop video conferencing (DVC) systems are relatively inexpensive. And the leading software packages used with DVC hardware are free for the downloading or can be purchased for a nominal cost. These software packages integrate other collaboration tasks into video face-to-face meeting features, such as the capability to share ideas using an electronic white board and to share documents.

Video conferencing and always-on DSL are a great match. If you have an Internet access account with public routable IP addresses, anyone on your LAN can use video conferencing for making or receiving video calls. The IP address (and a subdomain name) act as an always-on telephone number for receiving calls. A video conference participant calls the IP addresses using their video conferencing software to make a connection to the receiver. A message box pops up on the receiver's screen identifying the caller. You can choose to receive or disregard the call.

DVC system primer

Depending on the type of DVC system and how the vendor packages the product, the configuration of components varies. The three leading desktop video conferencing systems connect through the parallel (printer) port, Universal Serial Bus (USB) port, or PCI bus. The cameras that connect to a PCI video card offer the best performance. Using a USB DVC system means that you don't need to use a video card. Instead, the video camera plugs into the USB port on your computer. The quality of the pure software DVC solution inherent in the USB approach isn't as good as using a PCI video card, but it does offer the DVC option for notebook users.

A complete PCI DVC system includes the following elements:

✔ **A video/audio card.** The card includes ports for connecting the video camera and audio devices (microphone, headphones, speakers). The video is sent through your TCP/IP connection.

✔ **A small digital video camera.** This typically mounts on top of your monitor or has its own stand. Most video cards support the leading camera connections, including composite video through an RCA connector, S-VHS through a four-pin mini-DIN connector, and a Multimedia Extension Connector (MXC). The S-VHS port enables you to connect a camcorder to your video conferencing system. In most cases, an S-VHS camcorder offers a better quality image as well as other sophisticated camera features.

✔ **Video conferencing software.** Beyond the software to make the DVC system work (drivers and configuration software), you need a video conferencing software package. The two leading ones on the Internet are Microsoft NetMeeting and White Pine Software's CU-SeeMe.

Following are the leading video conferencing PCI and USB products:

- ✔ Supra Video Kit from Diamond Multimedia (www.diamondmm.com) uses a PCI video card.

- ✔ Videum Conference Pro (PCI) from Winnov (www.winnov.com) is available in both PCI and USB versions.

- ✔ C-it from Xirlink (www.xirlink.com) is a USB system.

- ✔ Bigpicture Video Phone from 3Com (www.3com.com) uses a PCI video card.

- ✔ Kodak DVC323 from Kodak (www.kodak.com) is a USB system.

- ✔ Internet Video Kit from Genius (www.genius-kye.com) is a PCI system.

- ✔ ZoomCam USB from Zoom Telephonics (www.zoomtel.com) is a USB DVC system.

Video conferencing software, such as CU-See-Me or Microsoft NetMeeting, uses dynamic TCP ports to deliver the data. Many firewalls, proxy servers, and routers don't support this feature, which means that the incoming video image is not allowed into the PC it is directed to. If you're using NAT on a router, you'll run into the same problem.

Most DVC systems are based on international communication and conferencing standards, including the International Telecommunications Union (ITU) T.120 standard for multipoint data conferencing and the ITU H.323 standard for audio and video conferencing. The H.323 standard specifies the use of T.120 for data conferencing functionality, enabling audio, data, and video to be used together as part of a complete online conferencing system. Support for these standards ensures that you can call, connect, and communicate with people using compatible conferencing products from other vendors. Other important standards that come into play with video conferencing are H.255.0, H.245, G.711, and H.261.

Microsoft NetMeeting

You can't beat the price for Microsoft NetMeeting — it's free. You can choose to download NetMeeting as part of the Internet Explorer package or you can download it separately from the Microsoft Web site:

```
www.microsoft.com/msdownloads/
```

Microsoft NetMeeting includes the six most wanted collaboration tools (audio, video, file transfer, chat, document/application sharing, and whiteboard). Figure 8-2 shows the Microsoft NetMeeting window. Video and audio are limited to only two conferees, but you can switch from one conferee to another on-the-fly. The audio and video quality are good over a DSL connection of at least 384 Kbps. NetMeeting is H.323 compliant to allow it to work with other video conferencing packages.

Figure 8-2:
The
Microsoft
NetMeeting
window.

CU-SeeMe

CU-SeeMe, at $69, is a video conferencing system that isn't as full-featured as Microsoft NetMeeting. However, the video and audio performance with CU-SeeMe are strong and easily equal the levels you get with NetMeeting. CU-SeeMe includes some collaboration tools, such as a whiteboard, a file transfer utility, and a text-based chat option, but lacks document sharing. An even more glaring problem with CU-SeeMe is that it isn't fully H.323 compliant. You can't make a direct connection to anyone using NetMeeting without going through a MeetingPoint Server from White Pine Software.

What is cool about CU-SeeMe is that you can create your own multipoint conferences. Instead of being limited to one-on-one video conferencing, you can put up to 12 video windows on the screen simultaneously, each displaying the face of a different conferee.

WebCams and video clips

You can leverage your video conferencing system as a WebCam or to create and send video clips through e-mail. After you have a video conferencing system set up on a PC, you can add WebCam software. A WebCam operates as a video surveillance camera pointed at something, such as an office. The

WebCam software takes images from the video camera at designated intervals, such as every few seconds, and uploads them to a Web site. Web site visitors can then view the images, which can be refreshed automatically or by clicking the Refresh or Reload button.

You can get WebCam software for $25. One of the best programs is WebCam32, which you can buy online and download from the following site:

```
www.kolban/webcam32/
```

This shareware product is relatively easy to set up and use. Figure 8-3 shows the WebCam32 program window.

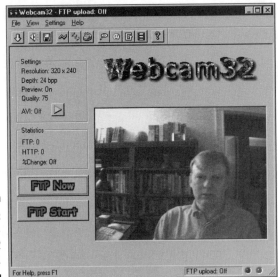

Figure 8-3:
A Web-
Cam32
window.

You can use a program such as VDOMail (`www.vdo.net`) to capture and send video clips using your e-mail client. You send the video clip inside its own runtime program so that those receiving the clip don't need to be running a special program on their site.

Net Phones, Faxes, and Pagers

Net phones, faxes, and pagers bring popular communications tools to the Internet. Net phones and video conferencing are closely related, and in some cases are combined into a single application, such as Microsoft NetMeeting. Net phone applications allow you to make and receive voice calls over the Internet.

Net-based faxing uses software and a fax provider service to route faxes to and from the Internet to fax machines connected to the PSTN or as e-mail messages to other Net users. IP paging uses a simple IP pager program running on your system and other users' systems on the Internet, and acts like the pagers used in telecommunications.

IP phone home

Voice over IP, VoIP for short, forms the basis of Net phones. Like a standard telephone, Net phones let you send your voice anywhere in the world. Whereas the telephone network carries your voice as an analog signal over copper wires, using Net phones transforms your voice into digital packets. These packets are sent over the Internet (or any IP network) and reassembled into one voice stream on the other end.

Traditional phone networks use a connection shared by only the two participants. Net phones, on the other hand, send their digital packets over the Internet, which is used to transmit lots of other data. On an IP network, no one knows when or whether pieces of data will arrive on the receiving end. When they are late, you experience delays in conversation — the person on the other end won't hear your sentence until all the packets have arrived. When packets don't arrive, you experience dropouts, or missing words or phrases.

The big appeal of Net phones is simple: You save money on telephone calls. The big problem with Net phones is that they don't sound as good as regular phone calls. You experience delays between when you speak and when the other party hears you, and the other party's voice will sound thin and shaky.

The use of codecs (compression/decompression algorithms) contributes to the poor sound quality of Net calls. Codecs squash data into its smallest possible form so that it will travel quickly. But because packets lose data during compression and decompression, codecs make voices sound weak and thin.

Here are the leading Net phone programs:

- ✔ Internet Phone (www.vocaltec.com)
- ✔ Net2Phone (www.net2phone.com)
- ✔ NetMeeting (www.microsoft.com)
- ✔ PhoneFree (www.phonefree.com)
- ✔ VDOPhone (www.vdo.net)

In 1996, the International Telecommunications Union created the H.323 standard, a specification that can make a Net phone work with any other H.323-compliant application. The drawback is that when two different applications talk, the sound quality deteriorates because the proprietary codecs that a company develops produce better-quality transmissions when both ends of the connection use the same Net phone program. A NetMeeting-to-NetMeeting call, for example, sounds better than a NetMeeting-to-VDO Phone call. However, a new voice-over-IP interoperability specification is emerging called iNow that will allow Net phone callers to talk with anyone regardless of the Net phone software they're using.

Net faxing

Mail-to-fax gateways give you the ability to send faxes as you would e-mail. The software you need to send faxes is part of the faxing service you decide to use. Several fax server companies are available, and you should shop carefully.

The Net faxing option may be a practical solution if you do a lot of faxes, but its feasibility for you depends on where the remote printing of the faxed document can take place. The role of the Net fax service is to make the links between the Internet and the PSTN send the faxes to machines connected — not to the Internet — to a POTS line. The Internet-to-PSTN links are not as prolific as POTS lines to the global PSTN.

Here are the leading Net fax providers:

- ✔ Faxaway (www.faxaway.com)
- ✔ FaxSav (www.faxsav.com)
- ✔ Faxscape (www.faxscape.com)
- ✔ FaxStorm (www.netcentric.com)
- ✔ JFAX (www.jafax.com)

Buzz me (IP paging)

A DSL connection enables the effective use of IP paging. IP paging does two basic things: shows who's online and enables you to flash messages back and forth. The IP pager detects whether the person you want to page is connected to the Internet, and if the person is, the IP pager sends the person a real-time alert.

Typically, the IP pager launches at computer start-up and stays active, listening for incoming alerts or waiting for you to beep someone. This is where the beauty of the always-on DSL connection comes into play. With the IP pager listening on your PC, you can be instantly paged by anyone, anywhere on the Internet. The IP pagers ping each other at short intervals to see who's online. If the ping is successful, the client reports that your buddy is online and ready to chat. Unfortunately, no two IP paging programs can talk to each other, so you need to go with the market leader or make sure everyone you want to page is using the same IP pager software.

The leading IP pager program by far is ICQ (pronounced "I seek you") from ICQ, Inc., which is owned by America Online. ICQ lets you find friends and associates online in real time and communicate with them directly. Figure 8-4 shows the ICQ program window.

Figure 8-4:
The ICQ
window.

ICQ includes a broad selection of communication tools, including chat (both one-on-one and multiparty), file transfer, and URL sharing. The client also comes with several important privacy features, such as the right to decide who can add you to their list. You can create a contact list containing information on all your online buddies. ICQ can automatically alert you about their online presence. You can also link ICQ to audio and video applications (such as Microsoft NetMeeting).

You can check out more about the ICQ program and its constellation of related paging services at the following address:

```
www.icq.com
```

You can download the program from

```
www.zdnet.com
```

or

```
www.download.com
```

Although you register for ICQ using your e-mail address, your IP address enables others to see and message you.

I Want My VPN

A *VPN (Virtual Private Network)* behaves like a private, secure network but runs across a public network such as the Internet. VPN is designed for tele-workers and companies with remote branches that want to communicate securely between two points. The marriage of DSL and VPN promises to open up secure networking for companies of any size.

VPN offers some compelling benefits for companies, including the following:

- ✓ **Save money**. Remote users bypass the telephone system for long-distance dial-up calls. Instead, users connect through their local ISP.

- ✓ **Bypass modem banks.** Connecting remote users through the Internet means that your company doesn't have to purchase and manage banks of modems with remote access server software to support dial-in users. Instead, remote access users connect to your LAN through the same Internet connection your company uses to connect to the LAN. All data traffic comes through the DSL connection.

- ✓ **Add security for remote access connections.** VPN enables you to set up filtering to manage the traffic you allow to enter your local area network resources.

The VPN arena offers a host of software and hardware solutions. It remains an unsettled place, however, because several forms of VPN are floating around, including the following:

- ✓ **Microsoft's PPTP (Point-to-Point Tunneling Protocol).** This software protocol allows private communications to be carried over the Internet. PPTP lets you create a virtual private network using Microsoft Windows NT Server and Windows 95, 98, and NT clients. Using PPTP, you can have a secure link to your organization's network as if you were operating on the same local network. Microsoft provides the PPTP software as part of Microsoft Windows. PPTP is tied to Dial-Up Networking in Windows clients and Remote Access Service (RAS) in Windows NT as well as Microsoft's Windows NT routing solutions.

- ✓ **The IPSec protocol.** This protocol is part of IPv6 (it started out that way, but is now a separate effort; IPSec is available for Ipv4) and forms the basis for the next generation of VPN. This IP security standard covers authentication and optional encryption of IP traffic between routers, host-to-host, or any user-to-host, which is what VPN is about. IPSec adds the security elements missing in IPv4 to make TCP/IP a more secure conduit for VPN traffic. IPSec provides confidentiality and integrity to information transferred over IP networks through network-layer encryption and authentication. Microsoft is building IPSec into Windows 2000 (formally NT 5.0). IPSec operates in two modes: transport mode and tunnel mode. IPSec uses transport mode to provide end-to-end security between two hosts (systems). In tunnel mode, IPSec places an original IP packet in a new packet. You can use tunnel mode to set up an IPSec tunnel between two hosts as well as between a host and a security gateway. A security gateway can be a tunnel server, a router, a firewall, or a virtual private network device.

✔ **IETF SOCKS v5 security protocol standard.** This option enables any TCP/IP application to securely transverse existing SOCKS v4- or v5-based firewalls or servers. A key component to network security is a way to bridge the private (intranet) and public (Internet) networks securely to traverse a network firewall. SOCKS is an open protocol and industry standard that enables secure firewall transversal. IETF (Internet Engineering Task Force) established SOCKS as a framework for TCP applications to securely traverse a network firewall. SOCKS acts as a proxy at the session layer, rather than at the application layer, which means that it's application independent and applies security services on a generic session-by-session basis. SOCKS v5 is the latest version of SOCKS. SOCKS v4 is already widely deployed in existing firewall products. SOCKS v5 adds support for authentication and encryption, UDP proxying, DNS and IPv6 support.

One product from Sonic Systems, Sonic Wall VPN, includes built-in support for VPN. Using two Sonic Wall VPN boxes at each end of the connection creates a VPN in which all the traffic between the two sites is secure. The latest DSL routers from Netopia and FlowPoint also support VPN.

Practicing Safe Downloads

A *computer virus* is a program that spreads copies of itself throughout a computer or network by attaching those copies to host files (usually program files, known as executable files) or e-mail messages. Viruses typically perform a number of additional disruptive actions.

Viruses are everywhere on the Internet, and you need to protect your PC or LAN from the havoc they can create. Although proxy servers and firewalls deal with the dangers of TCP/IP networking, they are not effective deterrents against viruses.

Following are the main virus categories:

✔ **Boot sector viruses** infect the boot sector of a hard drive or diskette and may also infect the partition table or Master Boot Record. These viruses are usually workstation based rather than server based. Merely booting the computer is all that's necessary to activate a boot sector virus.

✔ **File infector viruses** typically infect application programs. These viruses often add themselves to the code of the host file, increasing its size enough for an antivirus utility to detect the change. These viruses can spread over a network and are activated when the host is executed.

✔ **Macro viruses** are miniature programs designed to run inside a particular application program. They typically replace the normal, benign action of a particular program command with an abnormal and destructive action.

Within each primary category, several other classifications exist. Time bomb viruses wait for a specified date and time before taking action. Stealth viruses take steps to conceal their presence. Polymorphic viruses change their code to an encrypted pattern when they're not executing. Multipartite viruses have characteristics of both file infectors and boot sector viruses. Logic bombs wait for a specific event to activate a virus.

Because the line between data and programs is blurring, you have to worry about data files as well. For example, suppose a user downloads a document file from an outside source. When the user opens the file, the data file may run a little program called a macro, which can cause damage to other data files that the user works with later. The problem becomes worse if users spend a lot of time accessing resources off the Internet because viewer or plug-in programs that let Web browsers automatically run certain kinds of files can potentially activate data files that may be infected with a virus.

You should make antivirus software part of your Internet security plan. Some good antivirus packages are available, such as Norton AntiVirus and McAfee VirusScan. Antivirus software performs the following functions:

✔ **Prevents a virus from infecting your system in the first place.** The software recognizes hundreds of viruses by maintaining a database of virus fingerprints, or signatures. It can scan new files that a user downloads from an external source or places into a diskette drive, and raise an alert before the user copies or executes the program.

✔ **Provides an early warning system.** Antivirus software can detect a virus that slips through your early warning system and quickly alert you, so that you can repair and disinfect the affected file before the virus spreads and forces your entire LAN to shut down.

✔ **Recovers from a virus attack.** Good antivirus software can clean, repair, and disinfect the computer, or at least tell you when to throw in the towel and restore from a clean backup.

Viruses can make your life miserable. When it comes to combating viruses, an ounce of prevention is worth a pound of cure.

Working by Remote Control

Remote control software enables a remote client computer to take control of another computer (called a *host*) through an Internet connection. The remote client takes full control of the host computer system as if the remote user were sitting at the host computer. Remote control over the Internet through TCP/IP allows you to access office computers from a local ISP connection anywhere on the planet. Because DSL is an always-on server, you can use remote control software to get access to a computer running the remote control software anytime. To use remote control software, the host computer must be accessible from the Internet through a registered (public) IP address.

The leading remote control software programs include Symantec's pcAnywhere, Compaq's Carbon Copy, and Netopia's Timbuktu Pro. Check them out at the following sites:

```
www.symantec.com
www.compaq.com/products/networking/software/carbon copy
www.netopia.com
```

Chapter 9

Checking Out DSL Providers

● ●

In This Chapter

▶ Getting DSL service from a data or packet CLEC

▶ Checking out the leading CLEC DSL deployments

▶ Finding ISPs offering CLEC-based DSL circuit service

▶ How ILECs are packaging DSL service

▶ DSL service offerings from major ILECs

● ●

CLECs and ILECs are scrambling to deploy DSL service, but they're taking different approaches. CLECs are targeting small to mid-sized businesses as well as Internet power users with multiuser and Internet server needs. ILECs are focused on the single-user customer interested in speeding up Web surfing and files. These differences in deployment, however, are becoming blurred as CLECs and ILECs compete with each other in both markets.

CLECs and DSL Service

Empowered by the Telecommunications Act of 1996, CLECs (Competitive Local Exchange Carriers) are offering DSL circuits at affordable prices to ISPs and in some cases directly to end users. CLECs began their DSL service deployments in earnest in 1998 and are aggressively rolling out DSL services, often beating ILECs to many markets. They will be offering DSL service in most major metropolitan areas by 1999.

You can get DSL from several different types of CLECs, but the data, or packet, CLECs are the most aggressive in deploying DSL technologies on a national scale. These CLECs are focused exclusively on data applications, and therefore deploy DSL more rapidly. They typically offer DSL services through ISP partners in most major metropolitan markets but might also sell directly to larger corporate customers.This section explains the DSL service offerings, deployments, and ISP partners for the three leading national DSL CLECs.

The three leading national CLEC DSL circuit providers are Covad Communications, NorthPoint Communications, and Rhythms NetConnections. These DSL circuit providers sell DSL connections wholesale to ISPs as well as directly to large corporate clients. The first round of DSL deployments by these CLECs are targeted at large metropolitan markets because these markets represent the largest concentrations of potential DSL customers.

DSL CLECs are primarily focused on small to mid-sized business and teleworker markets. The main flavor of DSL they're deploying is SDSL (Symmetrical DSL), but they also offer ADSL (Asymmetrical DSL) and IDSL (ISDN DSL). SDSL is a good choice for CLECs and their DSL customers because it offers symmetrical service and uses the 2B1Q line coding technique, which is a stable modulation scheme that quietly integrates with existing ILEC cable bundles. SDSL can be widely deployed without creating the kinds of technical problems caused by ADSL service.

CLECs are moving toward a two-tier DSL service deployment that offers a business-class service and an economy class. Business-class DSL Internet access includes Quality of Service guarantees at a premium guaranteed bandwidth. The economy class is a best effort service for less demanding Internet access accounts at lower prices.

Because most DSL circuits installed by CLECs are for other ISPs, you probably won't have direct contact with the CLEC itself, except during the installation of the DSL service at your premises. The person who does the inside wiring for DSL service and sets up the DSL CPE (Customer Premises Equipment) for your ISP is almost always a CLEC employee or contractor.

The CLEC and ISP relationship

The CLEC and ISP relationship is one of a partnership between a wholesaler and retailer. The CLEC handles the DSL circuit (the telecommunications realm), and the ISP provides Internet access service bundled with DSL service.

CLECs offer DSL circuits to ISPs at a flat rate. Many ISPs offering DSL service resell DSL circuit services from more than one CLEC as well as from ILECs. The cost to the ISP for the DSL connection varies depending on the CLEC, the speed of the service, and other factors. The ISP sells the DSL circuit as part of an Internet service package. ISPs may break down the cost of the DSL circuit and the Internet service, but the prices they show for the DSL circuit aren't necessarily what they are being charged. The ISP might include a markup. You can't know what an ISP is paying for the DSL circuit because it's a negotiated deal.

Because the ISP determines the price for DSL service, prices vary from one ISP to another and from one market to another. The wide variances in DSL service pricing means that you can end up spending hundreds of dollars more per year for the same DSL service from the same CLEC but using a different ISP. Pricing for DSL service depends on a variety of configuration and speed options. The foremost factor in pricing is speed. More speed costs more money. However, the more bandwidth you order, the better the value because one-time setup charges typically remain fixed regardless of the speed you choose within a specific DSL flavor.

One thing to keep in mind about differences in DSL service pricing is that an ISP might be more expensive but offer better service in terms of a lower number of subscribers contending for the same Internet access bandwidth. Unfortunately, verifying ISP claims about subscribers to bandwidth ratios is almost impossible for customers.

In their efforts to gain market share and new DSL customers, DSL CLECs often run installation specials that waive the DSL circuit installation charges. Check the ISP Web sites for any specials being offered in your area.

Checking out CLEC DSL

Although the ISP is your interface to getting specific information about your DSL connection to the Internet, the CLEC's Web site is the best starting point for checking DSL availability in your area. The CLEC's Web site also provides links to their ISP partners, which you can then use to check out the ISPs servicing your area to do comparative shopping. If you know the ISPs in your area, you can also check out their sites to see whether they're offering DSL service.

CLECs and CPE options

Like most DSL providers, the CLEC DSL provider determines the DSL CPE that can be used with their DSL service by virtue of the DSLAM equipment they use at the CO. Because most CLEC DSL offerings are targeted at businesses and more-sophisticated Internet users, their CPE offerings include multiuser bridges, routers, and brouters. They support DSL CPE from computer networking companies such as Ascend, Cisco, 3Com, Netopia, FlowPoint, Cayman, and Efficient Networks.

The DSL CPE — and the DSL circuit — are sold through the ISP. And like DSL service pricing, costs of DSL equipment vary from one ISP to another. Most ISPs, however, try to keep CPE prices low because they represent one of the biggest start-up costs in getting DSL service. Pricing for your DSL CPE typically ranges between $300 to $900 for multiuser bridges and routers. Bridges typically cost considerably less than routers. Using DSL multiuser bridges and routers enables more IP options for your DSL Internet service, such as the capability to run Web servers and use IP video conferencing or voice applications.

NorthPoint Communications

NorthPoint Communications is a data CLEC focused on delivering business-class DSL services to small and mid-sized businesses and Internet power users. NorthPoint has been aggressively deploying and pricing their DSL service, with several price cuts since they began offering DSL circuit service.

DSL service options

NorthPoint's DSL data transport services are priced at a flat rate to ISPs based on the speed of the connection, which includes 160 Kbps, 416 Kbps, 784 Kbps, 1.04 Mbps, and 1.5 Mbps. Your DSL service cost will reflect the NorthPoint circuit costs plus the ISP Internet access service charges and ISP markup. NorthPoint brands their DSL service as NorthPoint DSL.

NorthPoint provides ISPs a Quality of Service (QoS) agreement called Service Level Agreements (SLAs). These agreements deal with congestion protection and latency. NorthPoint guarantees 99.99% packet delivery over the NorthPoint network to the ISP. This baseline ensures that a customer's data traffic will be delivered reliably from the end user through the NorthPoint network to the ISP's hand-off point. ISPs are packaging this service as a business-class service.

NorthPoint also commits to a round-trip latency of 10 milliseconds (ms). The round trip covers transmission of the packet from the subscriber's DSL CPE through the CO DSLAM to the multiservice switch at the NorthPoint central site and back. Minimizing *latency,* the time it takes for a data packet to move across a network connection, is also a critical component of business-quality data transport. It ensures that a customer's data traffic moves with minimum delay — at least between their CPE and the ISP. Such latency assurances are essential when supporting network-sensitive applications such as IP voice and video conferencing.

DSL service availability

NorthPoint was early out of the DSL deployment gate. As of the end of 1998, NorthPoint DSL service was available in the following ten metropolitan areas:

Boston; Chicago; Dallas; Detroit; Houston; Los Angeles; New York; San Diego; San Francisco Bay Area (Silicon Valley); Washington, D.C.

And the survey says

NorthPoint Communications recently conducted a survey of 100 randomly selected NorthPoint DSL users. The survey was conducted by ConStat, an independent IT market research firm. All end users in the survey had been using NorthPoint DSL between one and ten months, and the majority of those surveyed were long-time Internet users. Here are some of the interesting results:

- ✔ 93% installed DSL in a work location (business or home office) and use it for e-mail, information searches on the Web, and transferring files.

- ✔ 47% are hosting their own Web sites using the DSL connection.

- ✔ 92% would recommend DSL service to their peers.

- ✔ 84% feel DSL makes them more productive.

- ✔ 47% upgraded to DSL from dial-up analog Internet access.

- ✔ 45% upgraded to DSL from ISDN.

By the end of 1999, NorthPoint Communications will extend its services to include 25 metropolitan areas. As a result of the planned 1999 expansion, it's estimated that 40 percent of all businesses in the United States will have access to NorthPoint DSL. More than 20 percent of all residences will also have access to NorthPoint DSL. NorthPoint DSL will be available also in the following metropolitan areas by the end of 1999:

> Atlanta, Austin, Baltimore, Cleveland, Denver, Miami/Fort Lauderdale, Tampa/St. Petersburg, Minneapolis/St. Paul, Philadelphia, Pittsburgh, Phoenix, Portland (Oregon), Raleigh-Durham, Seattle, St. Louis

CPE options

NorthPoint uses Copper Mountain (www.coppermountain.com) as its DSLAM equipment vendor, which in turn defines what DSL CPE can be used by NorthPoint. Each ISP has its own CPE preferences, and many support only a subset of the available CPE options supported by Copper Mountain DSLAMs. Currently, NorthPoint Communications supports the DSL CPE listed in Table 9-1, but they will undoubtedly add other products as they're certified to work with Copper Mountain DSLAMs.

Table 9-1   **DSL CPE Supported by NorthPoint's DSL Service**

Vendor	Product(s)	Description
Ascend (www.ascend.com)	Pipeline 50 Pipeline 75	Ascend is a leading internetworking products and telecommunications equipment company. Their ISDN routers work with IDSL (ISDN DSL). This enables ISDN customers to convert over to always-on IDSL service without buying new DSL CPE.
Copper Mountain (www.copper mountain.com)	Copper Rocket 201 SDSL Router Copper Rocket 201 Bridge	Both devices were developed by Copper mountain to work with Copper Mountain DSLAMs. These boxes are used as the basis for similar Netopia and 3Com CPE offerings.
FlowPoint (www.flowpoint.com)	FlowPoint 2200-16 SDSL Router FlowPoint 144 IDSL Router	FlowPoint is an internetworking CPE product company. Their FlowPoint SDSL router is widely used.
Netopia (www.netopia.com)	Netopia M-7100 LAN Modem (Bridge) Netopia R-7100 Router	Netopia is an internetworking products company offering a complete line of DSL CPE options. Their bridge and router solution is based on Copper Rocket technology.
Cayman Systems (www.cayman.com)	Cayman SDSL 1401 Router	Cayman Systems is an internetworking products company developing DSL CPE options.
3Com (www.3com.com)	3Com SDSL modem Office Connect Remote 811 ADSL Router Office Connect Remote 840 SDSL Router	3Com is the leader in networking and internetworking products for the PC market. 3Com and Copper Mountain have announced a strategic alliance for developing and marketing CPE. The 3Com DSL CPE suite uses Copper Mountain technology. NorthPoint Communications supports the 3Com line of DSL CPE. Both of these products integrate into the stackable Office Connect family of hubs and switches. The first member of this product line, the 3Com SDSL modem, is based on Copper Rocket 201. 3Com will also be offering the Office Connect Remote 840 SDSL router and the Office Connect Remote 811 ADSL router.

ISP partners

The NorthPoint Web site (www.northpointcom.com) is a great place to check to see whether you can get DSL service in your metropolitan area. The NorthPoint Web site displays a national map of metropolitan areas NorthPoint serves, but the deployment map doesn't break down the service availability to local communities. The NorthPoint Web site includes links to the Web sites of NorthPoint's ISP partners. You can also fill out a form to get a call from an ISP in your area. Table 9-2 provides the URLs for NorthPoint's ISP partners.

Table 9-2	NorthPoint's ISP Partners
ISP	*URL*
Boston	
@Work	www.work.com
Concentric	www.concentric.net
Epoch Internet	www.eni.net
Flashcom	www.flashcom.com
iCi	www.ici.net
Shore.Net	www.shore.net
Verio New England	mid-atlantic.verio.net
Chicago	
@Work	www.work.com
Concentric Network	www.concentric.net
Flashcom	www.flashcom.com
Interactive Network Systems	www.insnet.com
Novacon	www.novacon.net
ThoughtPort	www.thoughtport.net
XNet Information Systems	www.xnet.com
Los Angeles	
@Work	www.work.com
American Digital Network	www.adn.net
Concentric	www.concentric.net
Epoch Internet	www.eni.net
Flashcom	www.flashcom.com
InteleNet Communications	www.intelenet.net
Internet Express Network	www.ienet.com
InternetConnect	www.internetconnect.net

(continued)

Table 9-2 *(continued)*

ISP	URL
LinkOnline Network	www.linkonline.net
Verio Southern California	scal.verio.net
Vinet Communications Internet Services	www.vinet.com
New York	
@Work	www.work.com
1 Terabit.Net	www.terabit.net
Digital Telemedia	www.dti.net
Eclipse Internet	www.eclipse.net
Flashcom	www.flashcom.com
InfoHouse	www.infohouse.com
Intercom Online	www.intercom.net
MediaLog	www.medialoginc.com
Panix.com	www.panix.com
Tribeca Technologies	www.tribecatech.com
Ultracom	www.ultracom.net
San Diego	
American Digital Network	www.adn.net
Concentric Network	www.concentric.net
Epoch Internet	www.eni.net
Flashcom	www.flashcom.com
InteleNet Communications	www.intelenet.net
Internet Connect	www.internetconnect.net
Internet Express Network	www.ienet.com
LinkOnline Network	www.linkonline.net
Verio Southern California	socal.verio.net
Vinet Communications Internet Services	www.vinet.com
San Francisco	
@Work	www.work.com
best.com	www.best.com

ISP	URL
Brainstorm Networks	www.brainstorm.net
Concentric	www.concentric.net
Dspeed Networks	www.dspeed.net
Epoch Internet	www.eni.net
Flashcom	www.flashcom.com
Washington, D.C.	
@Work	www.work.com
Concentric	www.concentric.net
Digital Select	www.digitalselect.com
Flashcom	www.flashcom.com

Covad Communications

Covad Communications (www.covad.com) is leading the charge in offering affordable DSL service. Covad is focused on delivering business-class DSL services to small and mid-sized businesses and Internet power users. Covad is aggressively lowering DSL circuit charges to make their service affordable to an even larger market.

DSL service options

Covad offers SDSL, ADSL, and IDSL services under the TeleSpeed service mark. Table 9-3 lists Covad's current TeleSpeed services.

Table 9-3	Covad's TeleSpeed DSL Service Offerings
Service Option (Downstream/Upstream)	**Speeds**
TeleSpeed 144 (IDSL)	Up to 144 Kbps/144 Kbps
TeleSpeed 192 (SDSL)	Up to 192 Kbps/192 Kbps
TeleSpeed 384 (SDSL)	Up to 384 Kbps/384 Kbps
TeleSpeed 768 (SDSL)	Up to 768 Kbps/768 Kbps
TeleSpeed 1.1 (SDSL)	Up to 1.1 Mbps/1.1 Mbps
TeleSpeed 1.5 (ADSL)	Up to 1.5 Mbps/384 Kbps

DSL service availability

Covad provides blanket coverage in the metropolitan areas it serves. For example, Covad's deployment in the Boston metropolitan area was available in 53 communities by the end of 1998. It will add another 15 communities by the end of 1999.

Covad's Web site provides an excellent resource for checking specific DSL service availability by metropolitan area. You can check availability starting from a national service availability map through a metropolitan area down to a specific community. The listing includes communities wired for DSL as well as communities scheduled for DSL service. To narrow your search to your specific location, Covad lets you check a database of availability based on your phone number and address.

As of the end of 1998, Covad communications offered DSL service in the following six metropolitan areas:

Boston; Los Angeles; New York; San Francisco; Seattle; Washington, D.C.

Covad plans to add the following metropolitan areas in 1999:

Atlanta, Austin, Baltimore, Chicago, Dallas, Denver, Detroit, Miami, Minneapolis, Philadelphia, Phoenix, Portland (Oregon), Raleigh, San Diego, Sacramento (California)

CPE options

Covad uses Diamond Lane and Cisco DSLAMs at COs. The Covad Web site provides a list of DSL CPE Covad supports, which includes the following:

- ✔ Cisco 1604 router (IDSL)
- ✔ Cisco 700 series routers (IDSL)
- ✔ Cisco 675 SOHO/telecommuter DSL router
- ✔ Diamond Lane ADSL router
- ✔ Efficient Networks SpeedStream 5250 SDSL bridge
- ✔ FlowPoint 144 IDSL router
- ✔ FlowPoint 2200 SDSL router

ISP partners

The Covad Web site is a great place to begin your search of DSL ISPs in your area. In addition to offering detailed DSL circuit service, the Covad site provides a list of ISP partners complete with links to their Web sites. Table 9-4 lists Covad's IPS partners by metropolitan area as of the end of 1998.

Table 9-4	Covad's ISP Partners
ISP Partner	_URL_
Boston	
BicNet	www.bic.net
Concentric Network	www.concentric.net
CyberAccess	www.cybercom.net
Flashcom	www.flashcom.com
Internet Technologies Group	www.itg.net
MegaNet	www.meganet.net
Net1Plus	www.net1plus.com
NetWay	www.netway.com
Nii.net	www.nii.net
Shore.Net	www.shore.net
Xensei Internet Services	www.xensei.com
ZipLink Internet	www.ziplink.net
Los Angeles	
95 Net	www.95net.com
Agenda.net	www.agenda.net
Caprica Internet Services	www.caprica.com
Cari.Net	www.cari.net
ClubNet	www.clubnet.net
CNM Network	www.cnmnetwork.com
Concentric	www.concentric.net
CTSNet	www.cts.net
DigiLink Internet Services	www.digilink.net
DSL Networks	www.dslnetworks.com
Epoch Internet	www.eni.com
Flashcom	www.flashcom.com

(continued)

Table 9-4 *(continued)*

ISP Partner	URL
Global Pacific	www.globalpac.com
GUS Network America	www.gus.com
IE Net	www.ienet.com
Infospec.net	www.ispec.net
Intelenet Communications	www.intelenet.net
Internet Express	www.ixpress.com
Internet Specialties West	www.iswest.com
InternetConnect	www.internetconnect.net
InterWorld Communications	www.interworld.net
NetQuest	www.netquest.net
Netroplex Internet Services	www.netroplex.com
Netsol.net	www.netsol.net
Netway Communications	www.nwc.net
NetWizards	www.netwiz.net
Packet Central	www.packetcentral.com
SoftAware	www.software.com
Subnet	www.subnet.org
Ultimate Internet Access	www.uia.net
ValueNet	www.value.net
Verio	Socal.verio.net
West Coast Internet	www.thewestcoast.net
XtraFast.com	Xtrafast.com
New York	
1 TeraBit.Net	www.terabit.net
Concentric	www.concentric.com
Connect2 Internet Networks	www.con2.com
DukeTech	duketech.com
Eclipse Internet Access	www.eclipse.net
Flashcom	www.flashcom.com
Frontline Communications	www.fcc.net
FullWave	www.fullwave.net
Garden Networks	www.garden.net

ISP Partner	URL
Globix	www.globix.com
i-2000 Internet	www.i-2000.com
IBS Interactive	www.interactive.net
InfoHouse	www.infohouse.com
Lightening.Net	www.lightening.net
MediaLog	www.medialoginc.com
North American Internet Service	www.nais.com
Simlab	www.simlab.net
SuperLink	www.njxdsl.com
Wired Business	www.wiredbusiness.com
San Francisco	
BJT	www.bjt.net
Bluetrain Networks	www.bluetrain.com
Brainstorm Networks	www.brainstorm.net
Concentric	www.concentric.net
Direct Network Access	www.dnai.net
DSL Networks	www.dslnetworks.com
Dspeed Networks	www.dspeed.net
Emf.net	www.emf.net
Epoch Internet	www.eni.net
Flashcom	www.flashcom.com
HyperSurf Internet Services	www.hypersurf.com
IDIOM Internet Services	www.idiom.com
iHighway	www.ihighway.net
Interstice	www.interstice.com
ionix	www.ionix.com
ISP Networks	www.isp.net
I-Step Communications	www.istep.com
LMi.net	www.lmi.net
MasterLink	www.m-l.net
MCE	www.mountaincomputers.com
Meer.net	w3.meer.net
NanoSpace	www.nanospace.com
Netmagic	www.netmagic.net

(continued)

Table 9-4 *(continued)*

ISP Partner	*URL*
NetWizards	www.netwiz.net
Network Architects	www.archnet.com
Sirus	www.sirus.com
SLIP.NET	www.slip.net
South Valley Internet	www.garlic.com
TransBay.Net	www.transbay.net
Tri Valley Internet	www.trivalley.com
Tycho Networks	www.tycho.net
ValueNet	www.value.net
Verio	www.verio.net
WebNexus	www.webnexus.com
Whole Earth Networks	www.wenet.net
Winterlink	www.winterlink.net
Seattle	
Flashcom	www.flashcom.com
Mpire.net	www.mpire.net
Speakeasy Network	www.speakeasy.net
Washington, D.C.	
Atlantech Online	www.atlantech.net
BusinessDSL.net	www.businessdsl.com
CAIS Internet	www.cais.net
Coretel	www.coretel.net
DigiNet Communications	www.diginetusa.net
Digital Select	www.digitalselect.net
DSL Express	www.dslexpress.net
Flashcom	www.flashcom.com
LTS	www.ltsweb.net
PatriotNet	www.patriot.net
The Hub Internet Services	www.knight-hub.com

Rhythms NetConnections

Rhythms NetConnections (`www.rhythms.net`) is focused primarily on working with large companies to support teleworkers and branch offices. They also provide DSL circuit service to ISP partners to support the small business market.

CPE options

Rhythms uses DSLAMs from Cisco, Copper Mountain, and Paradyne. Rhythms offers RADSL, SDSL, and IDSL services and support the following DSL CPE:

- Cisco (`www.cisco.com`) Cisco 675 and ISDN routers
- Copper Mountain (`www.coppermountain.com`) Copper Rocket 301 SDSL Router and Copper Rocket 201 IDSL Router
- FlowPoint (`www.flowpoint.com`) 144 ISDL Router and 2000 SDSL Router
- Netopia (`www.netopia.com`) R7100 SDSL Router and R3100 ISDL Router
- Paradyne (`www.paradyne.com`) 5446 and 5446 RADSL bridge and router plus MVL modem

DSL service availability

Rhythms plans to roll out service to 300 COs in the largest 35 metropolitan areas in the United States Rhythms' goal is to service more than 70 percent of the businesses and remote workers in any given service area. Rhythms currently offers DSL-based network services in the following four metropolitan areas:

Chicago, Los Angeles, San Diego, San Francisco

By the end of 1999, Rhythms will offer DSL service in the following metropolitan areas:

Atlanta; Austin; Baltimore; Boston; Cincinnati; Cleveland; Columbus; Dallas; Denver; Detroit; Hartford; Houston; Kansas City; Miami; Milwaukee; Minneapolis; New York; Philadelphia; Phoenix; Pittsburgh; Portland (Oregon); Raleigh; Sacramento; St. Louis; Salt Lake City; Seattle; Tampa; Washington, D.C.

ISP partners

The Rhythms Web site provides a listing of ISP partners for small businesses. Table 9-5 lists the ISP partners by metropolitan area, as of the end of 1998.

Table 9-5	Rhythms' ISP Partners
ISP Partner	*URL*
San Diego	
CariNet	www.cari.net
CONNECTnet	www.connectnet.com
CTS Network Services	www.cts.net
Internet Express	www.ixpress.com
K-Online	www.k-online.com
MilleniaNet	www.millennianet.net
SelectNet	www.select.net
Simply Internet	www.inetworld.net
zNet	www.znet.com
San Francisco	
Alameda Networks	www.alameda.net
Amer.net	www.amer.net
Creative.net	www.creative.net
DSP.Net	www.dsp.net
Nothing But Net	www.nothingbutthenet.net
Wombat Internet Guild	www.batnet.com

ISPs Acting as DSL CLECs

ISPs acting as their own DSL CLECs provide the DSL circuits along with the Internet service package. Their deployments tend to be localized. HarvardNet is an example of an ISP operating as a CLEC. They offer DSL service in the New England area based on Paradyne technology. Unfortunately, their DSL service is expensive compared to the DSL services from ISPs buying circuits from the CLECs. For example, a 384-Kbps ADSL connection costs between $499 to $1,926. Comparable SDSL service from an ISP using a DSL CLEC can range from $149 to $300. If DSL service isn't available from a CLEC, however, the HarvardNet solution is still a better deal than traditional T-1 service.

Finding a comprehensive listing of ISPs acting as their own DSL CLECs is not easy because there is no single information resource. You can, however, get a listing of all CLECs at the www.clec.com Web site. Unfortunately, the Web site doesn't separate DSL CLECs from voice and other types of CLECs.

ILECs and DSL Service

You know the ILECs as the telephone company that handles your voice and analog modem and ISDN connections. This section presents an overview of DSL service being offered by the following leading ILECs: Pacific Bell, US WEST, Bell Atlantic, BellSouth, Southwestern Bell, Ameritech, and GTE.

Most ILECs have aggressive deployment plans for their DSL services. Most ILEC DSL deployments will be to a much larger geographical area than CLEC offerings. According to TeleChoice (a telecommunications consulting and research firm), United States ILECs have deployed DSL at 788 COs as of the fourth quarter of 1998. These central offices service more than 17 million lines, making DSL fairly widely available going into 1999.

The primary DSL flavors deployed by the ILECs are ADSL, RADSL, and UADSL, which is based on the G.Lite standard. All these DSL flavors share the capability to be used over the same line as your POTS service. The ILECs are initially focusing on the single-computer, consumer market but some ILECs are offering DSL services targeted at the small business market as well.

ILECs are offering ADSL offerings through their own in-house ISPs as well as providing DSL circuit services to ISP partners. Most ILECs offer DSL circuits to ISPs and offer DSL Internet access through their own ISPs, which have names such as Bell Atlantic.net, PacBell.net, and USWest.net. If you get DSL service through an ILEC ISP partner, your DSL service might be billed separately by the ILEC. If you get DSL service from the ILEC ISP service, all charges are usually included on one phone bill.

ILEC offerings from their in-house ISPs are typically single-user solutions based on dynamic IP addressing and no support for DNS. In these types of accounts, you may need to buy proxy server software or an Ethernet-to-Ethernet router to share the DSL connection. The ILEC CPE options are typically single-user bridges, PCI DSL modem cards, or USB modems. Some ILECs, such as PacBell, are also offering business-class DSL services through their in-house ISPs.

In general, getting detailed, user-friendly information about ILEC DSL offerings from their Web sites can be difficult.

Pacific Bell

Pacific Bell, which is owned by SBC Communications, is one of the leaders in ILEC deployments of DSL service. PacBell is aggressively rolling out DSL service, driven by the fact that an estimated 35 percent of the nation's Internet traffic begins and ends in California. By the end of 1999, Pacific Bell plans to have the potential to provide ADSL service to 70 percent of its customers.

To get the latest information on the PacBell DSL offerings, including information on their ISP partners, check out the Pacific Bell Web site:

```
www.pacbell.com/products/business/fastrak/adsl/
```

DSL service options

Pacific Bell's DSL service offerings have been changing, with early 1999 price cuts for its DSL service. PacBell.net is now offering the following:

- **Consumer DSL Internet Access.** This package includes connection rates of up to 1.54 Mbps (guaranteed at 384 Kbps) downstream and 128 Kbps upstream. The cost is $49 a month for the Pacific Bell ADSL and Internet access (based on a one-year commitment). This is a single IP address and e-mail Internet access account that does not allow any commercial or business servers to operate on the DSL connection. If you commit to a one-year or longer contract, the DSL modem and service installation charge is waived. If you cancel before the contract ends, you are assessed an early termination fee of $125.

- **Enhanced DSL Internet Access.** This package includes connection rates of up to 6 Mbps (guaranteed at 1.5 Mbps) downstream and 384 Kbps upstream. The cost is $129 a month (based on a one-year commitment). The service includes five usable IP addresses and DNS service (for an additional fee of up to $200). If you commit to a one year or longer contract, the DSL installation and premises installation charge of $299 is waived. The DSL modem requires an NIC and costs $198. If you cancel before the contract ends, you are assessed an early termination fee of $125.

- **Business DSL Internet Access.** This package includes connection rates of up to 6 Mbps (guaranteed at 1.5 Mbps) downstream and 384 Kbps upstream. The cost is $328 a month for ADSL and Internet access (based on a one-year commitment). The service includes 29 usable IP addresses and DNS services. If you commit to a one-year or longer contract, the DSL installation and premises labor charge of $299 is waived. The DSL modem requires an NIC and costs $198. If you cancel before the contract ends, you are assessed an early termination fee of $125.

The Pacific Bell DSL service package may also be available from authorized ISPs. The ISPs offer the same $49 package as offered by PacBell.net, but some ISPs offer a better deal on the Enhanced Internet Access package. Check out the Pacific Bell ISP partners.

DSL service availability

Pacific Bell began deploying ADSL service in 1998 by offering DSL service in the following California communities:

Alameda, Albany, Alhambra, Anaheim, Arcadia, Berkeley, Beverly Hills, Bishop Ranch, Burbank, Burlingame, Canoga Park, Colma, Compton, Concord, Corona Del Mar, Costa Mesa, Culver City, Danville, El Toro, Escondido, Fair Oaks, Fremont, Fullerton, Garden Grove, Glendale, Hayward, Hollywood, Irvine, La Crescenta, La Jolla, La Mesa, Laguna Niguel, Livermore, Los Altos, Los Angeles, Milpitas, Mountain View, National City, Newport Beach, Northridge, North Hollywood, North Sacramento, Oakland, Palo Alto, Pasadena, Pleasanton, Redwood City, Reseda, Sacramento, San Bruno, San Carlos, San Diego, San Francisco, San Gabriel, San Jose, San Mateo, San Ramon, Santa Ana, Santa Clara, Sausalito, Sherman Oaks, Simi Valley, Sunnyvale, Tustin, Van Nuys, Ventura, Walnut Creek, West Hollywood

Pacific Bell announced an aggressive ADSL deployment for 1999, which will add the following California communities. Pacific Bell says this deployment will serve 70 percent of its customers.

Agoura, Antioch, Aptos, Arlington, Arroyo Grande, Auburn Mn, Bakersfield, Balboa, Benicia, Blue Revine, Boulder Creek, Brea, Brentwood, Buena Park, Carlsbad, Carmel, Chico, Chula Vista, Clayton, Clovis, Corona, Cotati, Davis, Del mar, Douglas, Edgewood, El Cajon, El Dorado, El Monte, Encinitas, Eureka Main, Fairfield, Fallbrook, Fresno, Gardena, Grass Valley, Half Moon Bay, Hawthorne, Hercules, Hesperian, Hollywood, Ignacio, La Brea, Lafayette, Larkspur, Lodi, Lomita, Martinez, Menlo Park, Merced, Mill Valley, Millbrae, Mission, Mission Viejo, Modesto, Monterey, Moraga, Msvl Franklin, Napa, Nevada City, Newhall, Nimbus, Oceanside, Orange, Orinda, Oroville, Pacific Beach, Pacifica, Palmdale, Paramount, Park Sorrento, Petaluma, Pittsburg, Placentia, Placerville, Poway Midland, Rancho, Redding, Richmond, Riverside, Rosemead, S. Placer Rocklin, San Clemente, San Juan, San Luis Obispo, San Marcos, San Pedro, San Rafael, Santa Cruz, Santa Marguerita, Santa Rosa, Scotts Valley, Sebastopol, Shingle Springs, So. Tahoe Sussex, Solamint, Sonoma, Stockton, Tiburon, Torrance, Tracy, Truckee, Turlock, Ukiah, Union City, Vacaville, Vallejo, Visalia, Vista, Watsonville, Woodland, Yorba Linda

ISP partners

Pacific Bell also provides DSL transport services to ISP partners, which are listed in Table 9-6.

Table 9-6	Pacific Bell's ISP Partners
ISP	*Web Site URL*
ArgoTech	www.argotech.net
Bay Area Internet Solutions	www.bayarea.net
Concentric	www.concentric.net
Direct Network Access	www.dnai.com
Flashcom	www.flashcom.com
In Reach	Business.inreach.com
JetLink	www.jetlink.net
JetNet	www.jet.net
Orconet	www.orconet.com
Pacific Bell Internet	www.pacbell.net
Sirus	www.sirus.com
SlipNet	www.slip.net
Znet	www.znet.com

US WEST

US WEST was out of the DSL starting gate early with a range of service options from $40 to $175 a month. US WEST offers DSL service with Internet access through !nterprise (www.interprise.com). MegaBit Services is the umbrella name for US WEST's family of DSL services. Table 9-7 describes the US WEST DSL service offerings based on a one-year contract. US WEST uses RADSL (Rate-adaptive Asymmetric DSL) as the basis of its DSL offerings. Pricing may change from state to state.

Customers are assessed a one-time installation charge and a monthly charge. You must also purchase a Cisco modem ($299) or PCI card ($199) from US WEST. Keep in mind that these prices don't include the Internet access charges.

US WEST sells the MegaLine, MegaOffice, and MegaBusiness DSL service through ISP partners.

Table 9-7		US WEST'S !nterprise DSL Offerings	
Service	*Speed*	*What It Is (Internet Access?)*	*One-Time Fee*
MegaLine	256 Kbps/ 256 Kbps	$40 a month with no Internet service	One-time connection fee of $110 plus a Cisco modem ($299) or PCI card ($199)
MegaOffice	512 Kbps/ 512 Kbps	$62 a month with no Internet service	One-time connection fee of $110 plus a Cisco modem ($299) or PCI card ($199)
MegaBusiness	768 Kbps/ 768 Kbps	$77 a month with no Internet service	One-time connection fee of $110 plus a Cisco modem ($299) or PCI card ($199)
MegaBit	1 Mbps/ 1 Mbps	$120 a month with no Internet service	One-time connection fee of $110 plus a Cisco modem ($299) or PCI card ($199)
MegaBit	4 Mbps/ 1 Mbps	$480 a month with no Internet service	One-time connection fee of $110 plus a Cisco modem ($299) or PCI card ($199)
MegaBit	7 Mbps/ 1 Mbps	$840 a month with no Internet service	One-time connection fee of $110 plus a Cisco modem ($299) or PCI card ($199)
MegaPak	256 Kbps/ 256 Kbps	$60 a month *including* Internet service	One-time connection fee of $400 plus a Cisco modem ($299) or PCI card ($199)

DSL service availability

US WEST is currently offering DSL service and partnering with ISPs in the states and communities listed in Table 9-8.

Table 9-8	US WEST's DSL Deployments
State	*Communities*
Arizona	Phoenix, Tucson
Colorado	Boulder, Colorado Springs, Denver, Fort Collins, Greeley
Idaho	Boise
Iowa	Ames, Cedar Rapids, Council Bluffs, Des Moines
Minnesota	Minneapolis, St. Cloud, St. Paul, Rochester
Montana	Helena

(continued)

Table 9-8 *(continued)*

State	Communities
Nebraska	Omaha
North Dakota	Fargo
Oregon	Albany, Corvallis, Eugene, Portland, Salem
South Dakota	Sioux Falls
Utah	Bountiful, Clearfield, Holladay, Kaysville, Kearns, Murray, Orem, Provo, Salt Lake City
Washington	Olympia, Seattle, Spokane, and Tacoma
Wyoming	Cheyenne

ISP partners

US WEST includes a listing of ISP partners offering DSL service using US WEST DSL service. The US WEST portion of your bill is added to your telephone bill. You can check out this list at www.uswest.com/interprise/dsl/isplist.html. The site lists over 100 ISPs offering US WEST DSL transport services.

Bell Atlantic

Bell Atlantic has been slow out of the DSL deployment starting gate but plans a big rollout of ADSL services in 1999. Bell Atlantic is offering ADSL service under the InfoSpeed brand name. Bell Atlantic will also be selling DSL transport services to ISPs. Check out the Bell Atlantic InfoSpeed Web site at

```
www.bell-atl.com/infospeed
```

for the latest DSL service information. Bell Atlantic is in the process of getting approval for a mega-merger with GTE, the large independent ILEC. The effect of this merger on their DSL deployments is unknown.

DSL service options

Bell Atlantic InfoSpeed service offerings are listed in Table 9-9. Current Bell Atlantic DSL product offerings through Bell Atlantic.net are single-computer DSL consumer solutions, with dynamic IP addressing and no DNS service. The Personal InfoSpeed and Professional InfoSpeed DSL offerings support only 90 Kbps upstream, which is one of the lowest upstream DSL offerings around. Bell Atlantic will also be offering an upgrade deal for its ISDN customers. Bell Atlantic is currently using a single-user bridge from Westell (www.westell.com).

Table 9-9	Bell Atlantic's InfoSpeed DSL Offerings	
Service	*Speed*	*What It Is*
Personal InfoSpeed	640 Kbps/90 Kbps	$59.95 a month with Internet service; $39.95 without. One-time connection fee of $99, plus a $325 ADSL modem, plus a $99 installation fee.
Professional InfoSpeed	1.54 Mbps/90 Kbps	$110 a month with Internet service; $60 without. One-time connection fee of $99, plus a $325 ADSL modem, plus a $99 installation fee.
Power InfoSpeed	7.1 Mbps/680 Kbps	$190 a month with Internet service; $110 without. One-time connection fee of $100, plus a $325 ADSL modem, plus a $99 installation fee.

On the surface, getting the Power InfoSpeed service, with 7.1 Mbps/680 Kbps for $190 a month with Internet service or $110 without, seems like a good solution for a LAN-to-Internet connection. Think again — this InfoSpeed service uses dynamic IP addressing and doesn't support DNS. If you want to share the DSL connection, you'll need to buy proxy server software or an Ethernet-to-Ethernet router.

Bell Atlantic and America Online (AOL) have a strategic alliance whereby Bell Atlantic will provide DSL service for AOL customers. AOL will be announcing DSL pricing in mid-1999, but the DSL upgrade is expected to cost AOL members less than $20, in addition to their AOL monthly costs.

DSL service availability

Bell Atlantic plans to service 2 million lines for ADSL in 1998 and 5 million lines in 1999. Table 9-10 lists the current Bell Atlantic DSL deployments by states and communities. As for other areas of the Bell Atlantic territory, little information is available for the following states: Delaware, New Jersey, Maryland, West Virginia, Connecticut, Maine, Massachusetts, New Hampshire, New York, Rhode Island, and Vermont.

Table 9-10	Bell Atlantic's DSL Deployments
State	*Communities*
Maryland	Bethesda, Beltsville, Colesville, Hyattsville, Landover, Rockville, Silver Spring, Suitland, Wheaton
New Jersey	Cliffside Park, Elizabeth, Englewood, Hackensack, Hoboken, Jersey City, Leonia, Newark, North Bergen, Oradell, Rutherford, Union City
Pittsburgh	Beaver Falls, Bethel Park, Carnegie, Connellsville, Glenshaw, Greensburg, New Castle, New Kensington, Oakland, Squirrel Hill, Uniontown, Washington
Philadelphia	Ardmore, BalaCynwyd, Bryn Mawr, Bethayres, Chestnut Hill, Coatesville, Collegeville, Downingtown, Huntington Valley, Jerkintown, Oaklane, Perkasie, Phoenixville, Royersford, Souderton, Willow Grove
Virginia	Alexandria, Annandale, Arlington, Baileys Crossroads, Falls Church, Merrifield, Vienna
Washington, D.C.	Dupont Circle, Georgetown, Northwest D.C.

BellSouth

BellSouth is a real sleeper when it comes to deploying its DSL service, FastAccess. As of the end of 1998, BellSouth was offering ADSL service only in the following areas:

Atlanta, Birmingham, Charlotte, Fort Lauderdale, Jacksonville, New Orleans, Raleigh

BellSouth claims it will add 23 cities in 1999. BellSouth services Alabama, Florida, Georgia, Kentucky, Louisiana, Mississippi, North Carolina, South Carolina, and Tennessee.

BellSouth is offering a 1.54-Mbps/256-Kbps link for $50 per month if you are a subscriber to the BellSouth Complete Choice or Business Choice telephone package. The cost for BellSouth FastAccess service as a stand-alone product is $60. Both offerings include Internet access from BellSouth.net as well as the ADSL circuit.

Customers will be charged a one-time installation fee for configuring their computer and phone line for FastAccess, which includes a $200 equipment charge for the FastAccess modem and related equipment and a $100 installation charge for the FastAccess line activation and on-site installation.

Unfortunately, BellSouth's Web site at

```
www.bellsouth.net/external/adsl/
```

isn't much help if you're trying to track down the specifics of DSL service.

Southwestern Bell

Southwestern Bell, which is owned by SBC Communications, has been a laggard in DSL deployment. In early 1999, however, it picked up the pace when it announced plans to deploy ADSL in 526 COs in its five-state region (which consists of Texas, Missouri, Oklahoma, Arkansas, and Kansas) by the end of 1999. Table 9-11 lists the scheduled FasTrak DSL deployments for Southwestern Bell in 1999. For more information about Southwestern Bell's DSL offerings, check out their Web site at

```
www.swbell.com/dsl
```

or call them at 888-SWB-DSL.

Table 9-11	Southwestern Bell's FasTrak DSL Deployments Scheduled for 1999
Metropolitan Area	*Scheduled Deployment Date*
Abilene	2nd quarter 1999
Austin	1st quarter 1999
Beaumont	1st quarter 1999
Dallas	1st quarter 1999
El Paso	2nd quarter 1999
Fort Worth	1st quarter 1999
Houston	1st quarter 1999
Kansas City	2nd quarter 1999
Little Rock	3rd quarter 1999
Lubbock	2nd quarter 1999
Oklahoma City	3rd quarter 1999
San Antonio	2nd quarter 1999
St. Louis	2nd quarter 1999
Topeka	3rd quarter 1999
Tulsa	3rd quarter 1999
Wichita	3rd quarter 1999

The Southwestern Bell ADSL offerings will be similar to Pacific Bell (both companies are owned by the same parent company, SBC Communications). These offerings may include the following (as presented by Pacific Bell):

- ✔ **A consumer DSL Internet access package.** This will include connection rates of up to 1.54 Mbps (guaranteed at 384 Kbps) downstream and 128 Kbps upstream. The cost is $49 a month for Southwestern Bell ADSL and Internet access (based on a one-year commitment). This is a single IP address and e-mail Internet access account that does not allow any commercial or business servers to operate on the DSL connection. If you commit to a one-year or longer contract, the DSL modem and service installation charge is waived. If you cancel before the contract ends, you might be assessed an early termination fee of $125.

- ✔ **An enhanced DSL Internet access package for multiple users.** This package includes connection rates up to 6 Mbps (guaranteed at 1.5 Mbps) downstream and 384 Kbps upstream. The cost is $129 a month (based on a one-year commitment). The service includes five usable IP addresses and DNS service (for an additional fee of up to $200). If you commit to a one-year or longer contract, the DSL installation and premises installation charge of $299 is waived. The DSL modem requires an NIC and costs $198. If you cancel before the contract ends, you might be assessed an early termination fee of $125.

Ameritech

Ameritech has remained quiet about DSL service deployments. SBC Communications, the parent company of Pacific Bell, Southwestern Bell, and other ILECs, is in the process of getting approval to add Ameritech to its stable. If that happens, Ameritech will probably be offering ADSL packages similar to Pacific Bell and Southwestern Bell.

Currently, Ameritech is offering its High-Speed Internet service only in Ann Arbor and Royal Oak Michigan, with deployments scheduled for the Detroit and Chicago areas. Ameritech.net's High-Speed Internet service offers 1 Mbps/128 Kbps for $60 a month. The service has a one-time installation fee of $150 plus a $199 charge for the DSL modem. The DSL service uses dynamic IP addressing, which means you can't have any DNS services associated with the account.

Unfortunately, the Ameritech Web site at

```
www.ameritech.com/products/data/adsl
```

offers little information for customers. To check out more on Ameritech DSL offerings, call 800-910-4369.

GTE

GTE is an independent ILEC offering telecommunications service in a number of states, including the following:

California, Florida, Hawaii, Illinois, Indiana, Kentucky, Missouri, North Carolina, Ohio, Oregon, Pennsylvania, Texas, Virginia, Washington, Wisconsin

GTE and Bell Atlantic will be merging, if they get government approvals.

DSL service options

GTE offers ADSL service through ISP partners and also from GTE.net. Table 9-12 lists GTE's DSL service offerings. These prices do not include Internet access. A one-time modem purchase of $300 is required, or you can rent the modem for $12 a month with a one-year or three-year service contract. GTE also charges a one-time connection fee of $60 plus an inside wiring and modem installation charge of $80. GTE DSL service uses an Orckit/Fujitsu bridge.

Table 9-12		GTE's DSL Service Offerings
Service	*Speed*	*What It Is*
DSL Bronze	256 Kbps/64 Kbps	$35 a month with no Internet service and with a 1-year contract. One-time connection fee of $60, plus an inside wiring and modem installation charge of $80, plus a modem purchase for $300 or a modem rental for $12 a month.
DSL Silver	384 Kbps /384 Kbps	$55 a month with no Internet service and with a 1-year contract. One-time connection fee of $60, plus an inside wiring and modem installation charge of $80, plus a modem purchase for $300 or a modem rental for $12 a month.
DSL Gold	768 Kbps /768 Kbps	$70 a month with no Internet service and with a 1-year contract. One-time connection fee of $60, plus an inside wiring and modem installation charge of $80, plus a modem purchase for $300 or a modem rental for $12 a month.

(continued)

Table 9-12 *(continued)*

Service	Speed	What It Is
DSL Platinum	1.5 Mbps/768 Kbps	$100 a month with no Internet service and with a 1-year contract. One-time connection fee of $60, plus an inside wiring and modem installation charge of $80, plus a modem purchase for $300 or a modem rental for $12 a month.
DSL Platinum Plus	1.5 Mbps/768 Kbps	Multiuser support. $230 a month with no Internet service and with a 1-year contract. One-time connection fee of $60, plus an inside wiring and modem installation charge of $80, plus a modem purchase for $300 or a modem rental for $12 a month.

DSL service availability

GTE is rolling out DSL in select cities across the United States. Table 9-13 lists GTE DSL deployments as of the first quarter of 1999.

Table 9-13	GTE's DSL Deployments Scheduled for 1999
State	**Cities**
California	Los Angeles, Palm Springs, Santa Barbara, Santa Monica
North Carolina	Durham
Florida	Tampa
Illinois	Bloomington
Indiana	Lafayette, Terre Haute, Elkhart, Ft. Wayne
Kentucky	Lexington
Missouri	Columbia
Oregon	Beaverton
Pennsylvania	Erie, York
Texas	Carrollton, College Station, Coppell, Dallas, Denton, Flower Mound, Garland, Grapevine, Irving, Lewisville, Plano, San Angelo, Texarkana
Virginia	Manassas, Dale City
Washington	Seattle, Everett, Redmond, Wenatachee
Wisconsin	Wausau

ISP partners

DSL Internet access must be obtained separately through one of the participating ISPs or from GTE.net. You can check out the GTE list of ISP partners at

```
www.gte.com/dsl/partisp.html
```

Chapter 10

Facing the ISP Interface

. . . .

In This Chapter

▶ Walking through a DSL Internet service installation process

▶ Going for a bridged or routed Internet service package

▶ Figuring out your IP addresses and DNS service needs

▶ Adding e-mail, network news, and Web hosting services

. . . .

*T*he DSL circuit is the pipeline for high-speed connectivity. What brings the DSL link to life are the value-added services of the Internet service provider. The ISP bundles the DSL circuit service with the IP addresses, Domain Name Services, and other TCP/IP-based services to define your Internet service.

What you can and can't do is defined in large part by the particular services offered by the ISP. This chapter takes you through the key elements that make up your Internet access package.

The DSL and ISP Connection

The United States has thousands of ISPs, and a wide variety of them offer DSL service. As a result, the packaging and pricing of DSL service is all over the map — literally and figuratively.

Only three types of ISPs, however, offer DSL-based Internet access service:

> ✔ **Independent ISPs.** These ISPs buy their DSL circuits from CLECs or ILECs and provide the bulk of DSL service offerings. In large metropolitan markets, independent ISPs typically buy DSL service from multiple DSL circuit providers. For example, an ISP serving Northern California might offer DSL service from Pacific Bell, NorthPoint, and Covad Communications.

✔ **ILEC ISPs.** These ILEC-owned ISPs provide an Internet access package added to the ILEC's DSL offerings. ILEC ISPs have names like PacBell.net, BellAtlantic.net, and US West.net. They compete with independent ISPs, but most focus on offering only dynamic IP address accounts targeted at the consumer market.

✔ **ISPs acting as CLECs.** These ISPs become CLECs by filing tariffs with the state regulatory agency, installing DSLAMs in COs, and using the ILECs' local loops in a similar way that larger CLECs do. One of the pioneers of this approach is Harvard Net, which is located in the Boston area.

Given the complexity of TCP/IP networking combined with DSL service issues, most ISPs could do a better job of creating user-friendly packaging of DSL-based Internet access services. ISP Web sites are the main source of initial customer information, yet many of these sites lack helpful information for customers trying to define their needs. In many cases, you have to call the ISP for basic service information. And even when you call, you may get a salesperson ill-prepared to answer your questions.

Walking through a DSL service installation

To install your DSL service, both the ISP and the DSL circuit provider perform many individual tasks. At several points during the provisioning process, an ISP may provide a status report on your installation (usually via e-mail).

The installation process for your DSL service can take from 10 to 30 days from the date you place your order. In some cases, the installation may take longer due to unanticipated problems with the phone circuits available to your location. The most common cause of delay is the presence of a load coil or repeater on your line. Such conditions require additional work to provision your circuit.

The following presents a step-by-step installation process scenario for getting DSL service up and running. Given the many ISPs and DSL circuit providers, your experience will vary. The following, however, will give you a general idea of what happens when you order DSL Internet access:

1. **You place the order with the ISP for your DSL service.**

    You order the specific data speed you want and the CPE to be used with your service as well as IP addresses, e-mail boxes, Domain Name Service, and any other service you want as part of your Internet access service. Keep in mind that the actual speed available to your premises may be less than what you want because of the distance of your premises to the CO.

At the time you order, you may have to provide a credit card number for the service. ISPs offering lower prices usually have the credit card as the preferred method of payment because it keeps their accounts receivable costs to a minimum. Your credit card isn't usually charged, however, until the service is installed and running.

2. **The ISP places a DSL circuit order with the DSL circuit provider and orders the DSL CPE.**

3. **The ISP receives a Firm Order Commit (FOC) date from the DSL circuit provider.**

   This is the date of the actual installation of DSL service. When the ISP receives the FOC date, it notifies you when the DSL circuit will be installed.

4. **The ISP registers or transfers your domain name and assigns IP addresses for your use, configures any secondary domain names, and configures your POP e-mail accounts.**

5. **The ISP records the order configuration information.**

   This information includes DSL circuit information, IP address assignments, default gateway, POP e-mail accounts and passwords, and domain name registration information in the order tracking system. The service order information sheet with this information is sent to you.

6. **The ISP configures its equipment to cross-connect your circuit to its network.**

7. **The ILEC completes a cross-connect of your circuit in the CO to the DSLAM equipment.**

8. **The ILEC installs the DSL circuit from the CO to the Minimum Point of Entry (MPOE) at your premises.**

   The MPOE is where phone lines first enter your facility. The MPOE can be a NID (network interface device) or an inside wiring closet. After the successful installation of the DSL circuit to the MPOE at your location, the on-site installation is scheduled.

9. **Within a few days following the installation of the circuit, a schedule is made for on-site installation and inside wiring.**

10. **On the scheduled on-site installation date, an installation technician arrives to install the DSL CPE and to complete the inside wiring between the MPOE and the DSL CPE.**

    The installation technician installs a new phone jack for the DSL circuit. If the DSL circuit will also carry normal voice telephone service, the installation technician installs a splitter to split the voice and data traffic. The existing inside wiring for your voice circuit is plugged into the splitter, as is the new inside wiring connecting to DSL CPE.

11. After the DSL CPE is installed and connected, the installation technician tests the circuit to ensure proper operation between the DSL CPE at your location and the DSL equipment in the CO.

12. After successfully testing the circuit, the installation technician installs a cable to connect the DSL CPE to your computer. The installation technician performs a ping test of the connection between your computer and the ISP's network. The DSL connection is up and running.

Navigating your CPE options

The DSL CPE available from an ISP is determined by the DSLAM used by the DSL circuit provider. The ISP selling the service sells the equipment supported by the DSL provider's DSLAM.

Your DSL CPE options fall into two distinct classes: CPE that supports only single-user service and CPE that supports multiple users. The IP configuration and DNS service options packaged with the Internet access service and the user policies of the ISP also define the capabilities of your Internet service.

Single-computer CPE consists of an external USB modem, an internal PCI adapter card modem, or a single-computer bridge. These configurations are typically dynamic IP address accounts with no DNS. In dynamic IP addressing, the ISP router acts as a DHCP server that dynamically assigns IP addresses based on different leasing times. Because your IP address changes randomly, you typically don't have a domain name, which means that you don't have DNS. This means you won't be running any kind of Internet servers or using IP video and voice applications. You can share these DSL connections, however, by using either an Ethernet-to-Ethernet router or proxy server software.

A multiuser DSL bridge or router allows you to share the DSL connection with computers connected to your LAN. The bridge makes your LAN part of the ISP's network and the Internet without any security; the router separates your LAN from the ISP's network and the Internet and adds some basic security protections. These two forms of CPE create the basis for either bridged or routed Internet access service.

Bridged DSL service

Multiuser bridged service is simple because the computers on your network are configured to be on the same IP subnet as the router at the ISP's side of the DSL circuit. No subnet routing is involved. The bridge doesn't include NAT or DHCP because it doesn't deal with filtering.

In a bridged service configuration, the ISP supplies IP addresses for each computer on your LAN that you want connected to the DSL service. The

Converting a single PC connection to a LAN connection

Many low-cost ILEC DSL offerings are targeted at single-user DSL consumers. For CPE, ILECs typically use PCI (Peripheral Component Interconnect) adapter cards and external single-user bridges as part of this DSL service. They also typically use dynamic IP addresses and don't support the use of DNS service.

The ISP router acts as a DHCP server and dynamically assigns IP addresses temporarily to your DSL CPE with different leasing times. Your IP address changes randomly, so you typically don't have a domain name. This means you don't have DNS, so you also won't be running any type of Internet server or using IP video and voice applications. You can share these DSL connections, however, by using either an Ethernet-to-Ethernet router or proxy server software.

A growing number of hardware and software solutions converts these single-computer DSL connections to multiuser connections. These products include Ethernet-to-Ethernet routers that sit between your LAN and the DSL bridge or software proxy servers. They allow you to share a dynamic IP address or a single static IP address account across multiple computers on a LAN. This allows you to share the DSL connection, but you'll need to use a Web and e-mail hosting service to use your domain name for a Web server and e-mail addresses.

downside of using a bridged service is that you have no firewall protection unless you purchase a proxy server or an Internet security appliance. As a bridge connection, you are part of the ISP's network with nothing protecting your network from their network. A firewall's purpose is to control access to your LAN, which it does by using identification information associated with TCP/IP packets to make a decision about whether to allow or deny access.

Routed DSL service

Routed service means that an IP router is installed between your local network and the DSL circuit. Routed DSL service uses subnet mask addresses. For computers on the LAN side of the router, two methods are possible for IP address assignments:

  ✔ **Using Network Address Translation (NAT).** NAT can be implemented on your router to conserve publicly routable IP address space. NAT allows the use of a single IP address on the router to make all inbound and outbound connections to and from the LAN computers. Most DSL routers support NAT. On the LAN side of the router, you can use private IP addresses that are not visible on the Internet. This provides an additional level of security by hiding your internal IP addresses. Using NAT and private IP addresses saves you the cost of leasing blocks of IP addresses but restricts the kinds of things you can do with your DSL connection because computers on your LAN don't have public, routable IP addresses. NAT is usually implemented with Dynamic Host Configuration Protocol (DHCP) to assign private IP addresses dynamically as needed. Because of the limited supply of publicly routable IP address space, many ISPs encourage the use of private IP address space and NAT DSL routed installations.

✔ **Using routable IP addresses.** IP addresses are typically leased in blocks of 8, 16, or 32. When you buy IP addresses from an ISP, the ISP ensures that those addresses are routable on the Internet. The ISP should also include custom subdomain names.

Putting Your IP Package Together

An essential part of your DSL service to the Internet is setting up your IP address and DNS services. If you want computers on your LAN to become hosts that can be accessed from the Internet, you need to use routable IP addresses. You'll typically assign the routable IP addresses statically to specific computers on your LAN to make them reachable at the same IP address.

These hosts will have full two-way communications over the Internet, which means that they can support such applications as IP voice and video conferencing or operate as a Web server, e-mail server, or any Internet-accessible server. Along with static IP addresses, you can assign secondary domain names to the hosts so that they can be accessed by using DNS.

Getting IP addresses

Blocks of IP addresses are available from most ISPs for a monthly cost based on the number of IP addresses. Some ISPs include a block of IP addresses as part of the service. The IP addresses assigned to you by the ISP are available for use while you are the ISP's DSL customer. They remain the property of the ISP and return to the ISP upon termination of the service.

Bridged or routed DSL service requires three IP addresses just for the IP server. One IP address is for the router, one is for the Ethernet connection, and one is for the WAN connection. If you get a block of eight IP addresses, for example, only five are available for hosts on your LAN.

For routed DSL service, you can get a block of 8, 16, or 32 IP addresses. These IP addresses typically cost around $25 a month per block of 8 addresses. You have to buy IP addresses in blocks of 8 due to the inherent characteristics of subnet routing. If you're using bridged DSL service, you can get IP addresses in any number because you don't use any subnet routing.

Domain Name Service

The ISP provides name resolution service so that TCP/IP applications can use Internet domain names instead of just numeric IP addresses. The ISP operates domain name servers, which your computers and applications can access for domain name resolution. It's very important to configure name resolution, because many applications depend on being able to use Internet domain names rather than IP addresses.

The DNS implements name-to-address assignments (forward DNS) and provides a lookup to determine a computer's IP address based on its domain name. The DNS also provides reverse mapping, to determine a computer's domain name from its IP address.

Make sure that your ISP provides entries in the DNS to implement reverse DNS assignments for your IP addresses and their assigned names. Many public servers on the Internet deny service to computers that do not have reverse DNS implemented, or whose name-to-address mapping does not match their address-to-name mapping. Your ISP should implement default values in its name servers for your IP addresses, for both name-to-address assignments and address-to-name assignments (forward and reverse DNS).

Often, failure of DNS name resolution is misinterpreted as a connectivity failure and a circuit outage, when in truth the application is simply not able to determine the IP address of the desired destination due to a problem with name resolution.

Getting your domain name

A domain name allows a company to maintain its own identity on the Internet—for example, `company.com`. This domain name can then become part of your Internet DNS address for e-mail addresses (`user@company.com`) and names of publicly accessible Web servers (`www.company.com`). You can use your domain name as part of your DSL service and Web and e-mail hosting services.

Before you can use a custom domain name for your company, you must register it with InterNIC (`http//:rs.internic.net`). Network Solutions operates InterNIC under contract to the National Science Foundation. They are responsible for administering domain name assignments within the `com`, `org`, and `edu` top-level domains. Domain names are registered on a first-come, first-served basis, although a flurry of activity has recently surrounded the issue of trademark protection with domain name assignments.

The steps for securing a domain name are basically as follows:

1. **Determine the availability of your desired domain name, by visiting** `http://rs.internic.net` **and using the Whois query tool.**

2. **After determining that your desired domain name is available, submit a registration request to the InterNIC or have the ISP submit the registration request on your behalf.**

   The registration can be accomplished online at the InterNIC Web site. You will need information, such as the addresses of the domain name servers that will support your domain.

3. **Proceed with plans to use the domain only after the InterNIC has confirmed the domain registration.**

   With the volume of domain registrations, frequently more than one person submits a request to register the same domain name.

4. **After the domain is registered, it must be configured on a name server.**

   The act of registering the domain with InterNIC does not mean that it is functional on the Internet. The domain must be configured on the name server(s) listed on the domain registration template submitted to InterNIC.

5. **The InterNIC invoices you directly for the domain registration fee.**

   Current costs are $70.00 for registering a domain name and the first year of use. After that, it's $35 a year. The ISP doesn't pay the domain name registration — you do.

Do not spend money creating business cards or marketing material that contains a domain name you have not yet secured.

Transferring your domain name from another ISP

If your domain was previously hosted at another ISP or hosting service and you want to move it to a new ISP's domain name servers, a few simple steps are needed to ensure a smooth transition. The domain transfer process can take place independent from your circuit installation process or other changes to your service. Follow these steps:

1. **Notify your previous ISP that you will be transferring your existing domain to your new DSL ISP's name servers.**

2. **Your DSL ISP will configure the domain on its name servers and duplicate the domain's zone information, including the contents of the domain.**

3. **When the domain configuration on the ISP's name servers is verified as operational, the ISP submits a domain modification transfer request to InterNIC.**

   The update causes the root or master name servers on the Internet to point to the ISP's name servers as responsible for your domain information.

4. **After InterNIC processes the update, the ISP's name servers are now authoritative sources of DNS information for your domain.**

   Your DSL ISP will notify the ISP that formerly hosted your domain that it (your former ISP) is no longer hosting the domain and should remove your domain information from its name server's configuration files.

Internet E-Mail Services

E-mail service is an essential component of any Internet connection. When it comes to getting Internet e-mail service, two options are typically available: hosting your e-mail services with your DSL ISP or operating your own mail server on your network.

Internet e-mail primer

Two primary electronic mail protocols are currently in use in the Internet world: POP (Post Office Protocol) mail and SMTP (Simple Mail Transfer Protocol) mail. Most ISPs support both protocols. POP mail client programs allow individual users to retrieve mail from their mailbox on the remote mail server. Most popular mail reading programs (such as Eudora, Netscape Messenger, and Microsoft Outlook) are POP mail client programs and can access a POP mailbox on a mail server.

SMTP is used on the Internet to deliver mail to a user's mailbox located on a mail server. An SMTP mail server handles mail for all user mailboxes at a given Internet domain, such as anyuser@company.com. The mailboxes for individual users are typically located on the SMTP mail server for the domain, whether it's located at an ISP or on a computer on your LAN. SMTP takes care of getting the e-mail messages to the mail server for your domain. Individual users must then have a way to access their mailboxes on the remote mail server. POP is the most common method of accessing mailboxes.

Most mail client programs deliver outbound mail by using SMTP. POP is used for picking up mail from your mailbox; SMTP is used for sending mail to others. If you have a mail server for your domain, your client computers typically should be configured to point to your mail server as the SMTP server to use for outbound mail. Otherwise they should be configured to send e-mail directly to the destination by using SMTP.

In general, configuring a POP client involves simply setting up your mail program so that it knows the following:

- ✔ **The IP address or name of the POP mail server where your mailbox resides**
- ✔ **Your username on the POP mail server**
- ✔ **Your password on the POP mail server**
- ✔ **The IP address or name of an SMTP mail server for sending outbound mail**

Most POP mail client programs have many more configurable user preferences. The basics of being able to access your mailbox, however, usually require only the information listed here.

Hosting your Internet mailboxes with the ISP

For smaller organizations, hosting your Internet mailboxes with the ISP makes the most sense. Each of your users has his or her own mailbox on the ISP's mail servers that uses your domain name as part of the e-mail address, such as david@angell.com. You can typically add e-mail boxes for a nominal fee of $2 to $10 per month.

Your users can use any POP mail client program to retrieve messages from their mailboxes. Mail from the Internet is delivered to the ISP's main server and then distributed to specific e-mail boxes.

ISP e-mail hosting frees you from the mail server administration and maintenance chores associated with running a mail server yourself. Users access their e-mail directly from their own e-mail box on the ISP server as well as from anywhere on the Internet. If you're using a dynamic IP Internet access account with no DNS support, you can use an e-mail hosting service tied to your domain name.

Running your own e-mail server

Installing and running your own SMTP mail server on your network allows you to manage your Internet e-mail along with your LAN e-mail. This management involves using an e-mail server program, such as Microsoft Exchange or MDaemon. If you elect to run your own SMTP mail server, mail for your domain is sent directly from the Internet to the mail server on your network. Connections are made directly from the mail sender on the Internet to your mail server and never touch any servers at the ISP.

Configuration and administration of the mail server is your responsibility. Typical configurations might include providing a mailbox for each user locally on the mail server, or implementing mail forwarding and redirection of a user's mail to a separate internal or external mail server where the user's mailbox actually resides.

Running a mail server requires that you have a domain name of your own registered with InterNIC and properly configured on the name server(s). The single most important requirement is that your name server be configured with an MX record for your domain. An MX, or Mail eXchanger, record tells everyone on the Internet which computer is acting as the mail server for your domain. Anytime someone sends e-mail to one of your users, the sending program looks up the MX record for your domain to determine where to deliver the mail.

Network News Service

Network news (also called *Usenet*) is a massive, distributed, text-based discussion group covering every topic imaginable. People who subscribe to newsgroups communicate by using a messaging system similar to e-mail.

Two components make up Usenet. First, thousands of news servers continuously pass articles back and forth as new articles are posted and old articles are flushed out. The news servers are to a large degree independent from each other, allowing a highly distributed and robust delivery mechanism. Second, individual users read and post articles from their local news server by using news client software. News client software is typically included as part of your e-mail or Web browsing program.

Your users may read and post Usenet news articles directly from the ISP's news servers by using the Network News Transfer Protocol (NNTP). This is the best way to go in most cases. Server-to-server Usenet news feeds are not provided with most DSL ISP service due to the sheer volume of news postings. Usenet news traffic currently averages between 10 to 15 GB per day. Delivering this amount of traffic over a DSL connection consumes all available bandwidth and makes the connection unusable for any other purpose.

Web Hosting

A DSL connection using routable IP addresses and Domain Name Service can support running a Web server on your LAN. However, evaluate the pros and cons of running your own Web server versus using a Web hosting service. Even if you use your domain name as part of your DSL Internet service package, you can use the www.yourcompany.com URL for a hosted Web server.

A Web site demands bandwidth, and if your site becomes popular, it can easily overload your DSL connection. In addition, you need to address a variety of administration and security issues in running a Web server on your LAN. Chapter 8 explains how to do some calculations for Web server bandwidth.

In many cases, you'll find it easier to use a Web hosting service instead of running your own Web server. Web hosting services are affordable, and you can easily manage them remotely by using a number of Web site management and content authoring tools, such as Macromedia's DreamWeaver or the Microsoft FrontPage program.

Part III
Connecting Your PC or LAN to DSL

The 5th Wave — By Rich Tennant

"I guess you could say this is the hub of our network."

In this part . . .

This part takes you to where the DSL connection meets your PC or LAN. For those who want to connect multiple PCs to your DSL connection, I tell you how to set up the basic LAN plumbing of network interface cards, cabling, and hubs or switches as the foundation for using a DSL bridge or router.

As part of making your PC or LAN connection to the Internet through DSL, you walk through configuring Microsoft TCP/IP for Windows 95/98 and Windows NT computers. If you're using a DSL bridge or router to connect to the Internet, you can configure the TCP/IP settings for your network interface card. If you're using a DSL modem card or a USB modem, you can find out how to work with Microsoft Windows Dial-Up Networking.

Chapter 11

Building Your LAN Foundation

*T*he high-speed, always-on nature of DSL service makes it ideal for local area network (LAN) to Internet connections. An Ethernet LAN enables your office or home to leverage the DSL connection by sharing it across multiple computers. Additionally, using Ethernet as the interface to a DSL connection through a bridge or a router makes Internet access transparent to you and other users. Simply double-click the Web browser (or any other TCP/IP application) and the Internet is instantly ready and waiting on your desktop.

Today, setting up the basic LAN plumbing of network interface cards, cabling, and hubs or switches is easy and inexpensive. This chapter explains how to assemble a LAN as the foundation for using a DSL bridge or router.

The LAN Plan

A LAN is a system for directly connecting computers so they can share resources. Networks are systems because they are made up of several components, such as cabling, hubs, network interface cards (NICs), and protocols. Ethernet is at the heart of PC-based LANs. Ethernet is a networking protocol based on the IEEE (Institute of Electrical and Electronics Engineers) 802.3 specification.

Building an Ethernet network from the ground up involves adding network interface cards (NICs) to your computers and connecting them with cabling to a network hub or a switch device. This forms the basic hardware infrastructure for your network and the foundation for connecting the DSL modem or router to your LAN.

After you have the LAN hardware in place, you configure the workstations for networking using the network operating system (NOS). Chapter 8 explains how to configure Microsoft Windows 95 and 98 for TCP/IP networking. Chapter 9 does the same for Microsoft Windows NT 4.0 (Server and Workstation).

Before moving on, you need to get some basic network concepts and terminology down:

- ✔ Networks are typically broken down into *workgroups,* which are clusters of computers that share resources among themselves. These workgroups can be comprised of an entire small network or a segment of a larger network.

- ✔ A *node* is a generic term for any device that can communicate with other devices on a network through a network interface. Nodes can be almost anything that can plug into a network — a workstation, a printer, a DSL bridge or router, a fax machine, and so on.

- ✔ A *workstation* is the general term given to any computer connected to a network. This term is often used interchangeably with *client.*

You can typically find all the networking hardware you need at your local computer retailer. However, check out some of the leading direct marketers of networking and data communications products for lower prices. You can find Data Comm Warehouse on the Web at www.warehouse.com or call them at 800-642-3064. Appendix A provides a listing of network and data communications product vendors.

Home networking through telephone lines

The Home Phoneline Networking Alliance (HomePNA) is an organization formed to develop specifications for interoperable, home-networked devices using telephone wiring already in place. The idea behind this initiative is to make networking within homes easier. The growth of online households and the growing number of homes with two or more PCs are driving home networking.

HomePNA's technology uses existing telephone wiring in homes as a shared medium for Ethernet networking. The first generation of HomePNA's networking specification can support up to 1 Mbps. A second generation of HomePNA's networking specification can support up to 10 Mbps. It remains to be seen whether this networking lite technology will catch on.

What's an Ethernet interface?

The Ethernet interface for Microsoft Windows is handled by the Network Driver Interface Specification (NDIS). Developed by Microsoft, NDIS provides a common set of rules for NIC manufacturers to use between the NIC and the operating system. NDIS provides compatibility between a NIC and the NOS that supports it. NDIS supports multiprotocol stacks. These stacks enable you to run different protocols concurrently on the same NIC.

Running multiprotocol stacks enables you to use the NIC for your LAN while at the same time using it to access another network. For example, you can run the Microsoft NetBEUI stack for your LAN and the TCP/IP stack for the Internet connection via a DSL bridge or router. Without the support for multiprotocol stacks provided by NDIS, you would have to unload one stack and then load another stack to access the network that uses a different protocol.

The LAN-to-DSL connection

Ethernet isn't just the preferred networking protocol for PCs. It's also the most popular interface for connecting DSL CPE to computers. DSL CPE devices using the Ethernet interface include bridges, routers, and brouters. DSL CPE is designed to support a LAN connect to a hub or a switch using standard networking cable used for Ethernet networks. Figure 11-1 shows how a DSL bridge or router connects to a LAN.

Figure 11-1:
A DSL bridge or router connects to the network hub or switch so that computers connected to the LAN can share the DSL connection.

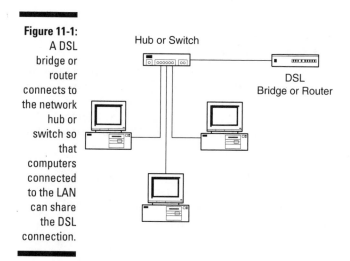

If it's connected, it has a MAC address

Each node on an Ethernet network has a unique address assigned to it called a MAC (Media Access Control) address. This is the protocol that controls access to the physical transmission medium on an Ethernet LAN. The MAC layer is implemented in an NIC or any other Ethernet device connected to the network, such as a DSL bridge or router. Every Ethernet device has a unique MAC address. An example of a MAC address is 00-80-AE-00-00-01.

MAC addresses play an important role in Ethernet networking and in the way DSL bridges work. Bridges work at the MAC layer, which means they work with MAC addresses to route data. Routers, on the other hand, work at the IP (Internet Protocol) layer to route data.

The number of computers supported by a given DSL bridge is controlled by the number of MAC addresses the bridge can recognize. A single-user DSL bridge recognizes data traffic from only the specific MAC address it was configured to accept. Multiuser bridges have a database of MAC addresses that they use to identify each Ethernet device on the LAN.

A typical multiuser DSL bridge learns all the MAC addresses of all the NICs on the local network. That way, the bridge can determine whether the data it receives is intended for the local network or for the Internet.

Single-user DSL bridges also use the Ethernet interface, but the bridge connects to a single computer's NIC instead of the hub. However, even these restrictive bridges can be tricked into supporting an entire LAN using proxy servers or Ethernet-to-Ethernet routers. These software and hardware products also act as firewalls to provide security from Internet intruders, as explained in Chapter 10.

Go fast with Fast Ethernet

Two Ethernet specifications are available for PC networks: standard Ethernet (called 10Base-T), which is based on the IEEE 802.3 standard, and the newer Fast Ethernet (called 100Base-T), which is based on the IEEE 802.3u standard. Fast Ethernet increases the speed capabilities of a network by a factor of 10 over standard Ethernet. Fast Ethernet supports up to 100 Mbps compared to 10 Mbps for standard Ethernet.

Fast Ethernet supports four types of media (cabling): 100Base-TX, 100Base-FX, 100Base-T4, and 100Base-T2. This chapter focuses on the most popular form: 100Base-TX, which operates over the same type of cabling used for 10Base-T and twisted-pair telephone cables.

Because both forms of Ethernet are based on the IEEE 802.3 standard, they're mutually compatible. This means you can integrate Fast Ethernet hardware into an existing 10Base-T network. In fact, 100Base-T cards actually support both 10 Mbps and 100 Mbps to accommodate mixed 10 Mbps and 100 Mbps networks. To use the greater speed, your network needs Fast Ethernet-compatible hubs and switches. Newer autosensing 10/100-Mbps hubs and switches can detect both forms of Ethernet and make adjustments on the fly.

The cost differential for 10Base-T versus 100Base-T networking hardware is rapidly becoming inconsequential. Going for the Fast Ethernet allows you to buy for your current networking needs and invest in a scalable technology for your future bandwidth demands.

An Ethernet network is a shared architecture that supports multiple computers and network resources running all kinds of applications. Today's multimedia applications demand far more bandwidth than their text predecessors did, and future applications will demand even more. Add multiple users all using bandwidth-hungry applications and you can get an idea of the demands a LAN faces. By adding DSL connectivity to your LAN, you enable users to do more things — now and in the future — on both the local network and the Internet.

Ethernet: Data bumper cars

Ethernet is based on the CSMA/CD protocol, which stands for Carrier Sense Multiple Access with Collision Detect. This long-winded term is comprised of a cluster of intertwined definitions that break down as follows:

✔ *Carrier Sense* means everyone attached to the network is always listening to the wire, and no one is allowed to send while someone else is sending. When a message moves across the wire, an electrical signal called a *carrier* is used. By listening to the wire, you know when it's busy because you sense the signal.

✔ *Multiple Access* means anyone attached to the network can send a message whenever he or she wants, as long as no carrier is being sensed.

✔ *Collision Detect* means if two or more senders begin sending at roughly the same time, sooner or later their messages collide on the wire. Ethernet hardware includes circuitry that recognizes when a signal has been messed up and immediately stops all sending; each sender must then wait a random amount of time before listening to the wire and trying to send again.

Because of the inherent limitations of CSMA/CD, the real-world capacity of an Ethernet network is only about 60% of the 10 Mbps or 100 Mbps numbers used for regular Ethernet or Fast Ethernet. So if you're using a 10-Mbps network, the capacity is more like 6 Mbps. Likewise, if you're using a 100-Mbps network, the real-world capacity is more like 60 Mbps. The bottom line on Ethernet networks is that performance degrades as more users are added to the LAN.

A star is born

In networking jargon, *topology* refers to the layout of the network. Every network type has a structure which is dictated to by the media it uses. Two basic network topologies are used for PC networks: *star* (hub and spoke) and *bus.* Although Ethernet supports both the bus and star topologies, the most popular by far is the star topology. In addition, Fast Ethernet works only with the star topology.

A star topology uses a hub or a switch as the center of the network. A *hub* is a simple device that acts as an interconnection point for the computers connected to the LAN. Each computer is connected to the hub using a cable that runs from the NIC to the hub. A switch plays a similar role as a hub except it includes more intelligence to better manage network data traffic.

Figure 11-2 shows how the hub sits in the center of a star network configuration. The cabling used for this network configuration is twisted-pair, and it uses RJ-45 connectors. The cables radiate from a central hub or switch to every computer or other network device on a network. The best feature of the star configuration is that damage to any given cable is likely to affect only a single computer, not the whole network.

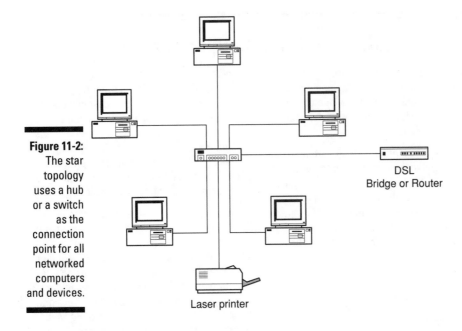

Figure 11-2:
The star topology uses a hub or a switch as the connection point for all networked computers and devices.

DSL
Bridge or Router

Laser printer

A Matter of Networking Style

Two basic types of network styles are used in the PC world: peer-to-peer and client/server. The network operating system (NOS) you're using defines which type of networking you can use. You can use Windows 95, 98, and NT Workstation to create a peer-to-peer network or to act as clients in a client/server network using Windows NT Server. These two forms of networking go beyond just the mechanics to reflect different styles in the way a network is administered and how users work on the network.

Nice and easy with peer-to-peer networking

A peer-to-peer network creates an environment in which all computers are created equal: Each computer on the LAN can act as a server, a client, or both. No dedicated computer controls operations; everyone is expected to be a good networker and share their resources (such as a program or a data file) with others. If you use Windows 95, 98, or NT Workstation 4.0, you have built-in peer-to-peer networking capabilities. For small networks, the peer-to-peer network is a reasonable network structure option, and you can always change your network structure to client/server as your needs dictate. The disadvantages of peer-to-peer networking center on the lack of centralized system administration for key network services, the lack of support for sophisticated multiuser applications, and minimal security options.

Power networking through client/server networking

A client/server network is made up of clients and a server that manages all kinds of services on behalf of the clients. These clients can include Windows 95, 98, NT Workstation 4.0, Mac, and UNIX workstations. Server functionality is added to the network by using a more powerful network operating system, such as Windows NT Server running on a dedicated server-class computer. The dedicated server enables more-sophisticated network management and security options, as well as a platform for more-sophisticated client/server applications.

By adding a server to your network, you make a new range of services available on your local network or to Internet users through your DSL connection. Run Windows NT Server on your LAN, and you can use Microsoft SQL Server to create sophisticated client/server databases that shares information across your LAN and the Internet. Using Microsoft Exchange Server, you can create your own in-house e-mail server. Using Internet Information Server, you can create a Web server. The downside of going the client/server route is that with the increased capabilities of client/server networking comes considerably more administrative complexity, as well as higher software costs.

NIC Your Computer

Any device that wants to communicate on a network must have a network interface. For connecting computers to an Ethernet LAN, you need a NIC (network interface card).

A NIC (pronounced "nick") is also referred to as a network controller. A network interface may come built into a computer, or you may need to install it yourself as an adapter card. The NIC connects to the cable that links your network together. Figure 11-3 shows a typical Fast Ethernet card.

NICs at work

In addition to providing the physical connection to the LAN, a NIC (pronounced nick) also performs other networking tasks.

A NIC prepares data so that it can go through the cable. The network card translates the data to go between the PC and the network cabling and back again.

A NIC addresses the data. Each network adapter card has its own unique address, which it imparts to the data stream. The card provides the data with an identifier when it goes out onto the network and enables data seeking a particular computer to know where to hop off the cable.

A NIC controls the data flow. The card contains RAM to help it pace the data so that it doesn't overwhelm the receiving computer or the cable.

A NIC makes and agrees on the connection to another computer. Before sending the data, the NIC starts an electronic dialog with the other PC on the network. The NICs agree on things like the maximum size of the data groups, the total amount of the data, the time interval between data chunks, the amount of time that elapses before confirming that the data has arrived successfully, and how much data each card can hold before it overflows.

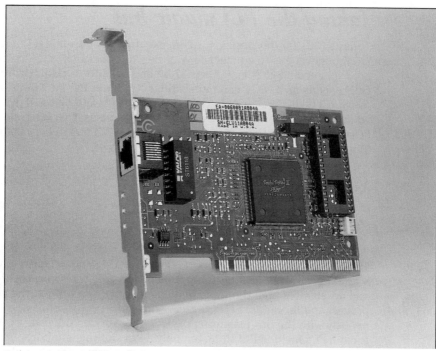

Figure 11-3:
A 3Com Fast
Ethernet
NIC.

(Courtesy of 3Com Corporation.)

Installing a NIC in a PC requires three steps:

1. Install the NIC in an available slot inside the PC. This part is painless.

2. Install a software driver so the OS will recognize the NIC. Driver software for most NICs is included with Windows 95, 98, and NT or provided on a disk that comes with the NIC.

3. Bind the networking protocol, such as TCP/IP, to the NIC. Chapter 8 explains how to bind and configure Microsoft Windows 95 and 98 for TCP/IP networking. Chapter 9 does the same for Microsoft Windows NT 4.0 (Server and Workstation).

You may choose from a cornucopia of NIC vendors to construct your LAN. Most NIC vendors typically offer a complete line of related network equipment such as hubs and switches. 3Com is the leader in NIC cards as well as most other networking products. Other companies offering popular network products include Intel, LINKSYS, and Bay Networks. Most Fast Ethernet cards are labeled as 10/100 cards and retail for under $100. Appendix A provides a detailed listing of NIC vendors.

Taking the PCI magic bus

Today's NICs for desktop computers use the PCI (Peripheral Component Interconnect) bus to connect to your PC. Notebooks use PC Card NICs, which are explained later. PCI supports 32-bit and 64-bit data paths, which dramatically speeds up a NIC's communications with your PC. Intel developed PCI to provide a high-speed data path between the CPU and up to 10 peripherals while coexisting with ISA and EISA expansion buses. Fast Ethernet NICs are PCI cards because they need the power of the PCI bus to handle the 100-Mbps capacity of Fast Ethernet.

There are other benefits using PCI network adapters besides LAN speed. The PCI data bus also provides a bus mastering technique that allows more processor independence. This technique reduces CPU overhead by taking control of the system bus, which enables the PC to support more bandwidth-intensive applications.

The PCI bus architecture supports bus mastering and concurrency. In *bus mastering,* an intelligent peripheral, such as a NIC, takes control of the bus and accelerates high-throughput, high-priority tasks without the need for processor intervention. *Concurrency* allows the processor to operate simultaneously with the bus mastering devices and work on other tasks. Most of today's PCs include PCI slots, but only a PCI slot that supports bus mastering can be used with a NIC that supports bus mastering. For example, 3Com's Fast Ethernet cards use a bus mastering architecture. In most PCs, slots 1 and 2 are bus mastering slots. If you're using a video PCI card, it's probably using one of the bus mastering slots.

Installing a NIC in a PC

The procedure for installing a network card in a PC typically involves opening the cover and inserting the NIC in an available PCI slot. After you install the NIC into the PC, the installation of the supporting software driver is handled by Windows 95, 98, or NT 4.0. The software driver allows an OS to communicate with the NIC.

The following instructions show you how to install a NIC into a PC. After the NIC is installed, go to the Windows 95 and 98 section or the NT section for specific instructions on installing and configuring the NIC's driver software.

Watch out for static electricity. Carry the NIC in the antistatic material it came in, and always ground yourself before you put your hands in a machine or handle computer hardware. To ground yourself to dissipate static buildup, simply touch the chassis of the computer before doing anything else.

Most NIC cards come with detailed instructions for installing the NIC and adding the driver.

Follow these steps to install a PCI Fast Ethernet adapter card:

1. **Turn off your PC and any peripheral equipment attached to it.**

2. **Remove your PC's case.**

   Each computer manufacturer has a different case design, but in general you remove the screws from the back of the computer case and then slide the case away from the chassis.

3. **Remove the back-panel cover for the free slot you want to use for the NIC.**

4. **Carefully slide the PCI network card into your PC's slot. Make sure all of the card's pins are touching the slot's contacts. Then secure the card's fastening tab to your PC's chassis with a mounting screw.**

5. **Replace your PC cover.**

6. **Connect one cable link for each PC.**

   Plug one RJ-45 connector into the network adapter card at the back of your PC, and plug the other connector into an available port on your hub.

7. **Turn on your PC and start Microsoft Windows.**

Installing a NIC driver in Windows 95 and 98

Installing NICs from Windows 95 and 98 is easy thanks to Plug-and-Play (PnP). Windows PnP automatically detects any new hardware you install. After you install the NIC in the computer and turn on the PC, Windows 95 and 98 detect the adapter card and prompt you for the software driver. Just follow the instructions on the screen to install the software driver.

You have two routes to installing the NIC driver. You can use the driver software provided by the NIC vendor or you can use the driver, if available, included with Windows 95 and 98.

Use the driver included with the NIC card because in many cases the NIC's driver is newer than the one included on the Windows 95 and 98 distribution media.

The following steps explain how to install a NIC driver in Windows 95 and 98 after installing the NIC in your computer:

1. **After you install the NIC and turn on your PC, Windows 95 and 98 detect the NIC, and the Select Device dialog box appears (see Figure 11-4).**

Figure 11-4:
The Select
Device
dialog box
shows all of
the NICs
that have
drivers
included
with
Windows 95
or 98.

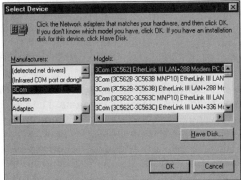

2. **If you want to use the driver from the disk provided with the NIC, insert the disk in the drive, and then click the Have Disk button. Or, in the Manufacturers list, select the vendor of the NIC; in the Models list, select the specific NIC and then click OK.**

3. **If Windows prompts you to insert the original Windows 95 or 98 distribution media, do so.**

   If the Windows 95 or 98 operating system files are already installed on your computer, Windows automatically finds the files and completes the driver installation.

4. **After the installation is complete, restart Windows 95 or 98.**

Installing a NIC driver in Windows NT

Installing a NIC driver in Windows NT 4.0 is handled differently than in Windows 95 or 98 because of the lack of Plug-and-Play support in Windows NT. However, the NIC installation software usually adds a PnP utility to support the NIC.

When you start Windows NT 4.0 after installing the NIC, there is no hardware
detection and no Add New Hardware Wizard. You install a NIC driver in
Windows NT using the Network properties in the Control Panel. To install a
NIC driver in Windows NT, do the following:

1. **After you install the NIC, double-click the Network icon in the Control
   Panel.**

   The Network dialog box appears (see Figure 11-5).

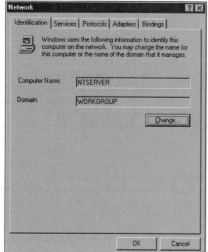

Figure 11-5:
The Network
dialog box.

2. **Click the Adapters tab.**

3. **Click the Add button.**

   Setup prepares the network card choices, and then displays the Select
   Network Adapter dialog box (see Figure 11-6).

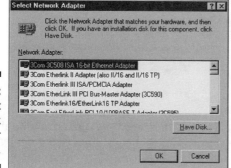

Figure 11-6:
The Select
Network
Adapter
dialog box.

4. **From the list box of Network Adapter cards, select the correct card to add to your workstation.**

   If you don't find your card on the list, insert the manufacturer's disk and click the Have Disk button.

5. **Click OK.**

6. **If Setup asks you to insert a disk or a Windows NT CD, do so and click OK.**

7. **Follow the directions on the screen. When you're finished, click OK to close the Network dialog box.**

   A reminder appears on-screen; you must restart your computer before changes take effect.

8. **Choose Restart to restart Windows NT.**

Connecting a notebook using a PC Card NIC

You can easily add a notebook computer to a LAN by using a PC Card (formerly called PCMCIA), which is a credit-card-size adapter card. You can also get dual-purpose Ethernet and analog modem PC Cards to support both networking and mobile communications. Figure 11-7 shows a LINKSYS Fast Ethernet PC Card.

Figure 11-7:
A LINKSYS Fast Enternet PC card.

(Courtesy of LINKSYS.)

If you have PC Card sockets (even if they're empty), when you double-click the PC Card (PCMCIA) icon in the Control Panel, you see the dialog box shown in Figure 11-8. PC Cards aren't automatically detected unless you first run the PC Card Wizard in Windows. To run the Wizard, double-click the PC Card (PCMCIA) icon in the Control Panel. If clicking the icon displays a Properties dialog box instead of the Wizard, PC Card support has already been enabled.

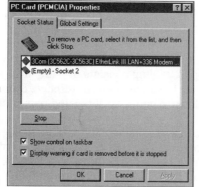

Figure 11-8:
The PC Card
(PCMCIA)
Properties
dialog box.

Windows 95 and 98 support PC Cards and run them in an enhanced Plug-and-Play mode to allow hot swapping. *Hot swapping* means Windows dynamically loads and unloads the proper drivers from memory when you switch PCMCIA cards. This means you can unplug a modem and plug in a network adapter without shutting down the system. Windows 95 and 98 recognize the change automatically, unload the software required for the previous card, and start the software you need for the network adapter.

Adding multiple NICs on a PC

Microsoft Windows 95, 98, and NT support the use of multiple NICs. You can connect any additional NICs in the same way you did the first NIC. The benefit of using two NICs and binding different networking protocols to each is that you can restrict any data traffic between the two protocols. For example, in the case of one NIC with TCP/IP and another for NetBEUI, access to the local network is restricted because the two protocols can't communicate. This creates a kind a firewall to protect your LAN from Internet intruders. For more information on using two NICs as a firewall, see Chapter 10.

The Media Is the Message

Cabling is the medium that connects your LAN. In the star topology, cabling runs from every computer (or other network device) to a hub or a switch. Network cabling used by most Ethernet networks is readily available, inexpensive, and easy to install. This section explains the fundamentals of network cabling.

Twisted pair isn't a deviant couple

Twisted-pair (TP) cabling is the predominant media used in the star topology for both 10Base-T and 100Base-T networks. Twisted-pair wiring is referred to as Category 5 cable, which gets its name from a cable rating system used by the Electronic Industry Association/Telecommunications Industry Association (EIA/TIA). Twisted-pair cabling consists of pairs of copper wires that are twisted around each other to cancel out interference. Ethernet uses four wires (two pairs) of the eight-wire capacity of twisted-pair cabling.

TP comes in two flavors: shielded twisted pair (STP) and unshielded twisted pair (UTP). UTP is by far the most popular. UTP consists of pairs of copper wires twisted around each other and covered by plastic insulation. STP has a foil or wire braid wrapped around the individual wires to provide better protection against electromagnetic interference. STP uses different connectors than UTP connectors, is more expensive than UTP, and requires careful grounding to work properly.

UTP Category 5 twisted-pair cable is readily available in retail computer stores and mail-order catalogs. You can buy it in a wide variety of lengths, such as 10, 25, and 50 feet, with male RJ-45 connectors at each end. Cable sold in specific lengths with connectors at each end are often labeled patch cables. The connectors at each end of the TP cable are RJ-45 male connectors that insert into female jacks in the NIC and the hub or switch port. You can use Category 5 data couplers to connect two UTP cables and extend the cable length. Many UTP cable vendors now offer color-coded cables that you can use to help identify which cable is connected to which device. For large cable installation jobs, you can buy cable in bulk and add connectors using a crimp tool.

Getting wired

If you're wiring a home or a small office, you can usually do it yourself. You need to get a single patch cable for each computer and any other network device you plan to connect to the LAN. You should also get extra cables as backup cables or for adding new devices.

If you want your network cabling installed inside walls or over drop ceilings, you may want to call in a cable contractor. Cable run through walls is terminated at a wall plate, much the way a telephone jack is installed. You need to use a fire-rated cable, which is more expensive than regular cabling.

Because labor is such a large part of network cabling costs, many businesses go ahead and wire every room or cubicle. The cost of doing the wiring when the installer is already there is less than it would be to bring the cable installer back to do the job a second time. Cable installation using a contractor generally costs five to ten times the cost of the actual wire, depending on the layout of your particular office.

Make sure you use someone with computer network cabling experience. Although electricians can lay the cable, they often lack the necessary network skills. They also typically lack the right test equipment to verify the integrity of the cable.

One-on-one with crossover link

If you're connecting a DSL modem or router directly to a single computer, you may need to use a special type of twisted-pair cable called a crossover cable. In a *crossover cable,* one of the four pairs is crossed to allow a LAN device to be connected directly to a single NIC instead of to the hub.

Cross-wired cables are usually designated with a color patch that differentiates them from other cables. If the crossover cable isn't marked, mark it so that it doesn't become mixed up with standard TP cables. You can check the connectors at each end to determine whether a TP cable is a crossover cable. If one of the wire pairs is crossed over at the connector, it's a crossover cable. Some DSL bridges and routers include special RJ-45 ports for connecting to a single computer using a typical fire-rated cable instead of a crossover cable.

Journey to the Center of the LAN

A hub or a switch is at the center of a LAN, with all network nodes — including your DSL bridge or router — connecting to it through cables. The fundamental difference between hubs and switches is one of intelligence. Hubs are passive devices that repeat all traffic to all ports. Switches can actively segment a network and forward packets to only the ports that need them, dramatically lowering traffic levels and increasing network performance. Hubs and switches can be connected together in different configurations to better manage a network or to enable the LAN to expand.

Most Fast Ethernet hubs and switches support both 10Base-T Ethernet operation and 100Base-TX operation. These devices are often referred to as *10/100
hubs or switches.* Each network device connects to a hub or a switch through
a port, which is a female RJ-45 connector. Hubs and switches come in many
port densities; port counts of 8, 12, 16, and 24 are the most common.

In addition to the variety of densities, most Ethernet hubs and switches are
stackable. *Stackable* hubs and switches allow an Ethernet LAN to start out
small and grow as needed because you can literally stack them on top of each
other to form a single network that has many hubs. These hubs use an
intrahub connection with a special uplink port. The connection between
stacked hubs carries network data and usually some control information.

Stackable hubs and switches can be linked using an intrahub connection.
Unfortunately, stackable hubs have one drawback: No standard exists for
connecting stackable hubs together. To be on the safe side, you may want to
use stackable hubs and switches that all come from the same vendor.

What's all the hub bub?

Hubs at their simplest level are simple, passive devices that keep network
data traffic moving. Hubs designed for small networks typically include four
to eight ports, and are about the size of a large external modem. Hubs have
LEDs on the front to tell you what's going on with the network traffic and
your node connections. Figure 11-9 shows a 3Com 10/100 hub.

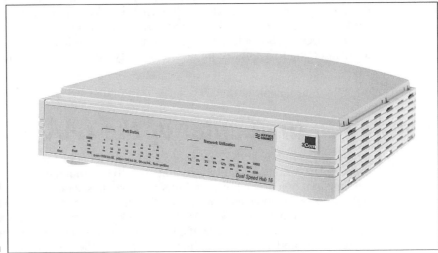

Figure 11-9:
A 3Com
10/100 hub.

(Courtesy of 3Com Corporation.)

Most hubs use one of the ports as an uplink port, which allows you to connect the hub to another hub or switch. Figure 11-10 shows how a hub forming the basis of one network can be linked to another hub to create a single, larger network.

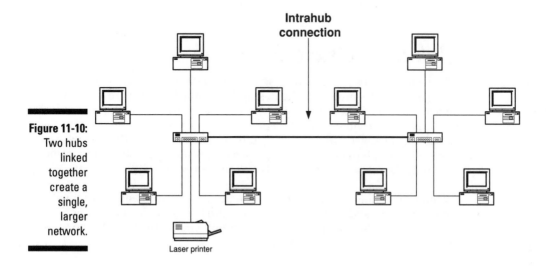

Figure 11-10:
Two hubs
linked
together
create a
single,
larger
network.

Hubs that support 100Base-TX also support 10Base-T are called autosensing or dual-speed hubs. These hubs adjust the ports to either 10Base-T or 100Base-TX, depending on the NIC connected to the port. Autosensing hubs are still more expensive than 10Base-T hubs, ranging in price from $200 to $700. However, those prices, like the prices of everything else in networking, are falling.

Intelligent hubs, or *managed hubs,* may also include network management software that lets you monitor hub status from a remote console. Typically, homes and small businesses with a single location use unmanaged hubs, which are less expensive. If you see the acronym SNMP, you're looking at a managed hub. With some hubs, you can buy management capability separately, so you may want to wait to see if you need it later.

If you're going for an autosensing hub that supports 100Base-TX and 10Base-T, you may want to consider going for a switch instead. Switches offer an attractive option for managing bandwidth for both 10Base-T and 100base-TX networks.

Without a switch, the maximum distance you can have between two hubs that are using Fast Ethernet is about 16 to 32 feet (5 to 10 meters). This can put a cramp in your networking options: If you're planning to connect two Fast Ethernet hubs together, make sure they're close enough together or use a switch between them. Fast Ethernet rules allow only two hubs to be connected within a single repeater domain.

Making the switch

When a network uses a hub, data sent by workstations is passed around the entire network, regardless of the destination of the data. This results in a lot of unnecessary traffic bouncing around the network, which can reduce the LAN's performance. A switch solves this problem because it listens to the network and automatically learns what workstations can be reached through what ports. The switch then selectively passes on data by transmitting the traffic from only the relevant port instead of all the ports, just like a hub does.

Switching technology has a big effect on network performance. Switches segment a network and provide extra bandwidth by managing data traffic. Switches are self-learning, so they learn about your network and keep up with any network configuration changes.

Switch workings

Each network adapter in your workstations has a MAC address, which is used to identify the workstation on the network. A switch might support a few or hundreds of MAC addresses.

Information is passed around the network as packets, and these packets contain MAC addresses. Using the MAC addresses, the switch can learn which workstations are connected to each of its ports. This information is stored in a switching database, which is a list containing each source address together with the port through which the device with that address is attached to the switch.

Using this database, the switch can act on packets it receives as follows:

- If the destination device is connected to the same port as the source device, the packet is discarded (called *filtering*), reducing unnecessary traffic.

- If the destination device is connected to a different port than the source device, the packet is transmitted through that port, which is called *forwarding*.

- If the destination device is unknown (not known in the switching database), the packet is transmitted through all other ports (called *flooding*).

Switches are more expensive than hubs, but the difference in price is getting smaller because switch prices are coming down. A good switch for a small network currently costs between $300 and $600. A number of switches designed for small networks come in a variety of port configurations. Like hubs, switches are usually stackable with other hubs and switches from the same vendor. Figure 11-11 shows a 3Com switch.

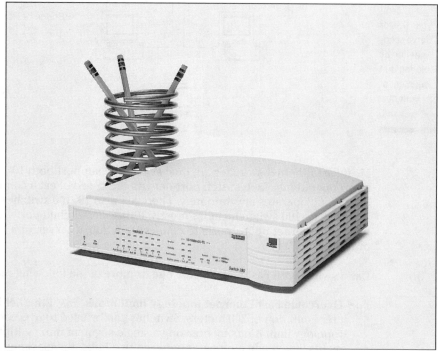

Figure 11-11:
A switch from the 3Com Office Connect family of networking products.

(Courtesy of 3Com Corporation)

You can connect hubs to a switch to increase the number of workstations connected to your network. These hubs can be either 10Base-T or Fast Ethernet hubs. By connecting your existing hubs to each other, you can leverage your investment and harness the power of switching. Figure 11-12 shows how you can use a switch with existing hubs and improve the performance of your server or DSL bridge or device.

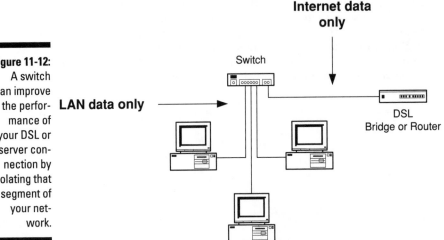

Internet data only

Switch

LAN data only

DSL
Bridge or Router

Figure 11-12:
A switch can improve the performance of your DSL or server connection by isolating that segment of your network.

Most Fast Ethernet switches are dual-speed and support both 10- and 100-Mbps operations. Each switch port detects the speed of each connected device and operates appropriately. There are also 10+100 switches. An Ethernet 10+100 switch includes mostly 10Base-T switched ports and one or two Fast Ethernet uplink ports to connect to a hub, a server, or a DSL bridge or router.

Use a switch with Fast Ethernet for one or more of the following reasons:

- ✓ **Overcome Fast Ethernet topology limitations.** Fast Ethernet has a diameter limitation of 200 meters. Switches can be used to overcome topology limitations by breaking a single segment into multiple segments. If you add a repeater between two hubs, the maximum network diameter jumps to 400 meters.

- ✓ **Improve performance by segmenting a Fast Ethernet network.** When a single segment gets too busy, users complain about response times, connections drop, or access to the Internet slows down. As the number of users on a segment increases or their demands on the network grow, network utilization drops. Using a switch can solve this problem.

- ✓ **Provide high network performance for specific nodes.** Any single node, such as a network server or a DSL bridge or router, can be connected to a switch that supports full duplex. Full-duplex operation allows information to be transmitted and received simultaneously and, in effect, doubles the potential throughput of the link. When you attach a switch with a full-duplex link to a node, you won't have any collisions when transmitting. In addition, the switch can send the node data anytime, not merely when the node is idle, because the link between the switch and the node consists of separate transmit and receive channels. This allows the network segment connecting the server or the DSL bridge or router to the LAN to be much more efficient. Not only is throughput doubled, but also the link can run at a much higher network utilization than a normal, half-duplex connection.

Chapter 12

Windows 95/98 Meets DSL

. . . .

In This Chapter

▶ Connecting DSL to Windows 95/98

▶ Setting up TCP/IP in Windows 95/98

▶ Configuring TCP/IP for LAN-based DSL connections

▶ Installing and configuring Windows 95/98 Dial-Up Networking

▶ Using Dial-Up Networking for DSL connections

▶ Working with some handy Windows 95/98 utilities

. . . .

*I*f your computer runs Windows 95/98 and you want to access the Internet through DSL, you need to install and configure the Microsoft TCP/IP stack. If you want to do the same from Windows NT, skip this chapter and read Chapter 13.

If you're using a DSL bridge or router to connect to the Internet, you need to configure the TCP/IP setting for your network interface card. If you're using a DSL modem card or a USB modem with Windows 95/98, you have to install the Microsoft TCP/IP stack, and you may also need to install and configure Windows 95/98's Dial-Up Networking. This chapter explains the fundamentals of configuring Windows 95/98 for connecting to your DSL CPE.

Windows 95/98 and DSL Connections

As explained in Chapter 5, you can use different types of DSL CPE as part of your Internet service. Various DSL CPE interface differently to Windows 95/98. If you use a DSL modem adapter card or an external USB (Universal Serial Bus) modem, you use Windows Dial-Up Networking to make the Internet connection through DSL. If you use an external DSL bridge or router, Windows 95/98 uses the Ethernet interface through a NIC (network interface card). Using an Ethernet solution means that you configure the NIC card for TCP/IP.

You can connect a Windows 95 or 98 computer to DSL in three ways:

- ✔ A DSL modem adapter card (PCI card)
- ✔ An external USB modem
- ✔ The Ethernet interface through a NIC

If you're using a DSL bridge or router, you install and configure the Microsoft TCP/IP stack for every Windows 95/98 computer on the LAN. You configure the TCP/IP properties of the NIC to enable your DSL connection.

If you're using a DSL PCI or a USB modem that uses DUN (Dial-Up Networking), it fools Windows 95/98 into thinking it is doing a dial-up connection, although DSL isn't a dial-up service. This means you'll still be using DUN in Windows 95/98 to work with either form of DSL CPE; the DSL connection is handled as a PPP (Point-to-Point Protocol), which is the standard protocol for dial-up connections to the Internet. A DSL modem can connect to your PC using PPP in two ways: PPP over ATM and PPP over Ethernet (PPPoE). See Chapter 3 for more information on these DSL PPP protocols.

When you install many PCI DSL modem cards or USB modems, they're treated as NICs in Windows 95/98. You install the software driver for the DSL modem using the same Network properties dialog box you use to install NIC cards. Other PCI DSL modem cards and USB modems are installed as modem drivers in the Modem Properties dialog box.

Check the power under the hood

Windows 95/98 requires certain minimum hardware capabilities to operate. Connecting to a high-speed DSL connection can also demand more from your PC. Higher bandwidth requires more PC power on several fronts, most notably the CPU, RAM, and video card. Here are some guidelines for the optimal configuration of a Windows 95/98 client computer:

- ✔ Use a Pentium or Pentium II class processor. More is always better when it comes to processor power. Most of the PCs sold today meet these requirements.

- ✔ Use an accelerated PCI bus video card that supports higher resolutions. Accelerated display cards make a difference, and the more onboard memory the display card has, the better. Larger monitors and higher resolution give you more real estate for multiple applications or larger viewing areas. Most new video cards easily surpass these minimum standards.

Setting Up TCP/IP in Windows 95/98

The first step in configuring Windows 95/98 is to install the Microsoft TCP/IP stack, if it's not already installed. If you've been connecting to the Internet through an analog or ISDN modem, Microsoft TCP/IP is probably running.

You can check to see whether the TCP/IP stack is installed by right-clicking Network Neighborhood icon, and then choosing Properties from the pop-up menu. The Network properties dialog box appears, as shown in Figure 12-1.

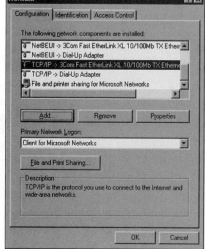

Figure 12-1: A NIC installed with TCP/IP listed in the Network properties dialog box.

In the Configuration tab, you should see the following entry:

TCP/IP ⇨ *network card adapter*

where *network card adapter* is the name of the installed network adapter card. If you see an entry like the following in Network properties, you already have the Microsoft TCP/IP stack installed:

TCP/IP ⇨ 3Com Fast EtherLink XL 10/100Mb TX Ethernet NIC (3C905B-TX)

If you don't have TCP/IP installed, you need to install the Microsoft TCP/IP stack. Installing the Microsoft TCP/IP stack binds TCP/IP to any installed NIC or DSL modem driver in the Network properties dialog box.

Before installing the Microsoft TCP/IP stack, have the original Windows 95/98 distribution media handy, in case Windows prompts you for it.

To install the Microsoft TCP/IP stack:

1. **Right-click the Network Neighborhood icon on your desktop and choose Properties from the pop-up menu.**

   Alternatively, you can double-click the Network icon in the Control Panel. The Network properties dialog box appears.

2. **In the Configuration tab, click the Add button.**

   The Select Network Component Type dialog box appears.

3. **Double-click Protocol.**

   The Select Network Protocol dialog box appears, as shown in Figure 12-2.

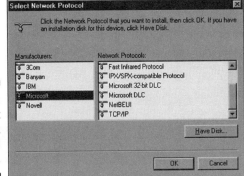

Figure 12-2:
The Select
Network
Protocol
dialog box.

4. **In the Manufacturers list at the left side of the dialog box, select Microsoft.**

   The available protocols appear in the Network Protocols list on the right side of the dialog box.

5. **Double-click the TCP/IP item.**

   When you return to the Network dialog box, scroll through the list of network components and you'll see TCP/IP listed. Notice that the component entry for TCP/IP uses the following notation:

   TCP/IP ➪ *network adapter name*

   TCP/IP is bound to your network adapter card.

6. **Follow the Windows prompts to complete the installation.**

   Windows 95 or 98 may prompt you for the original distribution CD-ROM or diskettes to complete the installation.

Configuring Windows 95/98 TCP/IP Clients

The next step towards DSL enlightenment is to configure your Windows 95/98 PCs for the type of IP network configuration you're using for your DSL connection. This process involves configuring the TCP/IP stack, which is bound to your NIC.

You can configure your TCP/IP stack in two ways:

✔ If you're using a DSL router as a DHCP (Dynamic Host Configuration Protocol) server to dynamically assign IP addresses, configure your Windows 95/98 clients as DHCP clients and don't add any specific IP address information to the TCP/IP properties.

✔ If you're using static IP addresses assigned to specific computers, manually configure the TCP/IP properties of each Windows 95/98 PC with IP address and DNS information.

Windows 95/98 allows only one TCP/IP stack configuration profile for LAN-to-Internet (or other network) connections. This means if you have more than one ISP connection available, you must change the settings in the TCP/IP Properties dialog box for each connection and then reboot the system. A shareware program called NetSwitcher (www.netswitcher.com) allows you to create multiple TCP/IP network connections and switch between them as needed. For example, if you're using a laptop and travel between different offices, each with its own LAN-based Internet access service, you can use NetSwitcher to create profiles for all of them and switch to the one you need when you're at a specific office. NetSwitcher also makes it easy to switch back and forth between a dedicated DSL connection and a dial-up connection when you're on the road.

Configuring Windows 95/98 clients for DHCP

Using a DSL router that acts as a DHCP server enables dynamic IP addressing. The DHCP server assigns an IP address from an available pool of IP addresses for use only during the current connection. The DHCP server functionality of a router allows IP addresses, subnet masks, and default gateway addresses to be assigned to computers on your LAN. You can use DHCP with registered IP addresses or private IP addresses along with the NAT (network address translation) feature incorporated in most DSL routers. Chapter 6 explains IP addressing and DHCP in more detail.

Windows 95/98's default TCP/IP stack configuration is as a DHCP client. A DSL router acting as a DHCP server will detect your PCs acting as DHCP clients. You may need to configure your DSL router to act as a DHCP server and tell it what IP addresses you plan to use.

If you've installed TCP/IP for the first time, you don't need to change any default settings to support DHCP. If you're unsure of your TCP/IP configuration, however, you can double-check to make sure the default TCP/IP settings are in place by performing the following steps:

1. **Right-click the Network Neighborhood icon on the desktop, and choose Properties from the pop-up menu.**

    The Network properties dialog box appears.

2. **In the network components list, select the network adapter with the TCP/IP binding, and then click the Properties button.**

    The TCP/IP Properties dialog box appears.

3. **Click the IP Address tab (shown in Figure 12-3) and select the Obtain an IP address automatically option (the default setting).**

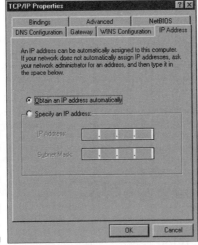

Figure 12-3:
The IP Address tab in the TCP/IP Properties dialog box.

4. **Click OK twice to exit the Network applet.**

5. **Restart Windows.**

Configuring Windows 95/98 clients for static IP addresses

If you plan on using registered IP addresses that you want to assign to specific computers — for example, to run a Web server — you need to configure each Windows 95/98 workstation with static IP address and DNS information. You can also use private (unregistered) IP addresses.

Before you configure your Windows 95/98 clients for static IP addressing, you need to get the following information from your ISP:

- ✔ An IP address for each Windows 95/98 workstation and any other Ethernet device. In most cases, these IP addresses will be registered IP addresses.

- ✔ A subnet mask IP address for your network.

- ✔ The IP address assigned to your DSL bridge or router, which is the gateway IP address.

- ✔ A host and domain name for the registered IP addresses.

- ✔ DNS server IP addresses, which are typically IP addresses supplied by the ISP.

Follow these steps to configure a Windows 95/98 workstation for static IP addressing:

1. **Right-click the Network Neighborhood icon on the desktop, and choose Properties from the pop-up menu.**

   The Network properties dialog box appears.

2. **Double-click TCP/IP⇨*network adapter name* in the network components list, or select it and then click the Properties button.**

   The TCP/IP Properties dialog box appears.

3. **In the IP Address tab, select the Specify an IP address option.**

   The IP Address and Subnet Mask fields become active.

4. **In the IP Address field, enter the IP address for the Windows 95/98 workstation you're configuring. In the Subnet Mask field, enter the subnet mask IP address.**

5. **Click the DNS Configuration tab, and then select the Enable DNS option.**

   All the controls in the Enable DNS group become active, as shown in Figure 12-4.

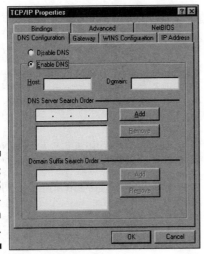

Figure 12-4:
The DNS
Config-
uration
tab.

6. **In the Host field, enter the host name for the Windows 95/98 worksta-tion uniquely associated with the specific IP address you entered in the IP Address tab.**

The host machine might be named david, which added to angell.com becomes david.angell.com.

7. **In the Domain field, enter your organization's domain name.**

For example, the host machine might be named david and the domain name might be angell.com.

8. **In the DNS Server Search Order area, enter the first DNS server IP address that you want your Windows 95/98 workstation to check, and then click the Add button. Do the same for each additional DNS server address you have.**

To remove any DNS server IP address from the list, select it and then click the Remove button.

9. **Click the Gateway tab, which is shown in Figure 12-5.**

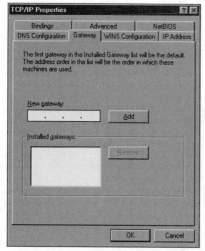

Figure 12-5:
The
Gateway
tab.

10. **In the New gateway field, enter the IP address of your DSL router, and then click the Add button.**

    If you're running multiple DSL connections and routers, you can add multiple gateway IP addresses. To remove any gateway entry from the list, select it and click the Remove button.

11. **Click OK twice to exit the TCP/IP Properties and Network dialog boxes.**

12. **If prompted, insert your distribution CD-ROM.**

    After installing the necessary information, Windows asks you to restart your computer.

13. **When promoted, click Yes to restart Windows with the new TCP/IP settings.**

Managing network card properties

Selecting the driver for your NIC in the Network properties dialog box and clicking the Properties button displays the properties dialog box for the selected NIC. This dialog box is similar to the one that appears when you use the Windows 95/98 Device Manager (the System icon in the Control Panel). A typical NIC displays three tabs in the properties dialog box for the selected NIC:

  ✔ The **Driver Type tab** displays the driver mode Windows 95/98 assigned for your NIC. The typical setting is Enhanced mode (32-bit and 16-bit) NDIS driver. This is the preferred mode in Windows 95/98 because it offers the best performance and the best memory management.

✔ The **Bindings tab** shows you all the protocols that are bound to the network adapter (see Figure 12-6). You can remove bound protocols by clicking on the check box. The best way to manage your protocols, however, is from the Installed network components list in the Network dialog box.

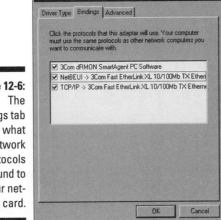

Figure 12-6:
The Bindings tab shows what network protocols are bound to your network card.

✔ The **Advanced tab** controls vary depending on the type of network adapter card. It shows you how the card is configured to handle traffic. Depending on your network adapter, the physical medium (cabling) may also be displayed in the Transceiver Type drop-down list. In most cases, you don't need to make any changes in this tab.

Managing your protocol bindings

Windows 95/98 supports multiple network adapters, and each of these cards can have different network protocols bound to it. For a given protocol to communicate with each network adapter on your computer, the network adapter driver must be bound to the protocol. *Binding* a protocol to a network adapter means you're attaching a software network interface to the network adapter. The binding defines the relationships between networking software components.

Whenever you install a new adapter card, Windows 95/98 automatically binds some protocols to it. You'll probably want to change the bindings — typically, you must add or remove bindings as your network changes.

You can change network adapter protocols by selecting the Bindings tab in the Adapter Card Properties dialog box. To display that dialog box, select the driver for the network adapter (the item in the list without a protocol name in front of it) in the Network properties dialog box, and then click the Properties button.

The general rule for binding protocols is to bind only the protocols that you need. Leaving unnecessary protocols bound to your network adapter slows down your network performance and can cause other problems. If you're not using it, remove it. You can add a removed protocol at any time.

Working the DUN Way

If you plan to use a DSL modem card or a USB modem that requires Dial-Up Networking, you need to install it. Before you can install DUN, however, you need to make sure that the DSL modem adapter card or USB modem is installed. When you install a DSL modem card or USB modem in Windows 95/98, the vendor's installation software installs the driver in the Network properties dialog box or the Modem Properties dialog boxes.

If you're using Windows 95, you need to use DUN version 1.2 or higher. Some DSL modem vendors include the latest version of DUN as part of their installation software. You can download the DUN upgrade from the Microsoft Web site (www.microsoft.com/msdownload/). If you're using Windows 98, you can use the DUN software that comes with Windows 98.

After you've installed DUN, you can create a connection profile. This DUN connection profile provides Windows 95/98 with the information it needs to make the DSL connection to the Internet. Some DSL modem vendors create the profile automatically as part of the installation. This installation software uses information added by the DSL Internet service provider.

The creation of a DUN connection profile mimics the creation of a DUN profile for an analog or ISDN modem — with one major difference. The entry for the telephone number is a virtual circuit (or channel) identifier (VCI) number. This 16-bit number is provided by the DSL ISP and identifies the channel your DSL service will use to make the connection.

Installing Dial-Up Networking

You may need to install DUN from your Windows 95/98 distribution CD (or disks). You can check to see whether DUN is installed by opening the My Computer folder on your desktop. If you see an icon labeled Dial-Up Networking, DUN is installed. If you don't see that icon, you must install the Dial-Up Networking software from your original Windows 95/98 disks or CD-ROM, as follows:

1. **Double-click the Add/Remove Programs icon in the Control Panel.**

   The Add/Remove Programs Properties dialog box appears.

2. **Click the Windows Setup tab.**

3. **Double-click the Communications option.**

   The Communications dialog box appears, displaying the available communications and networking components of Windows 95/98.

4. **Click Dial-Up Networking in the Communications dialog box (see Figure 12-7) and then click OK.**

   This returns you to the Windows Setup page.

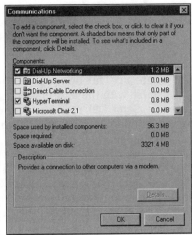

Figure 12-7: The Communications dialog box displays which Windows 95/98 components are installed and which aren't.

5. **Click OK.**

6. **Follow the Windows 95/98 prompts for installing the disks or CD-ROM that Windows needs to install Dial-Up Networking.**

7. **When prompted, restart Windows 95/98.**

Setting up a new connection profile

You can create a Dial-Up Networking profile by using either the Make New Connection icon in the Dial-Up Networking window or by using the Internet Connection wizard. To create a connection using the Make New Connection icon in the Dial-Up Networking window, do the following:

1. **Double-click the My Computer icon on the desktop, double-click the Dial-Up Networking folder, and then double-click the Make New Connection icon.**

   The Make New Connection wizard appears.

2. **Click Next.**

3. **Type a name (such as DSL) to identify your connection, and then select your DSL device from the list of modems.**

   This list includes any USB modems as well as any modem cards that you've installed.

4. **Click Next.**

   A page appears for you to type the telephone number of the host computer that you want to call.

5. **Type the Virtual Circuit Identifier number, and then click Next.**

   This 16-bit number is provided by your DSL ISP.

6. **Click Finish.**

Configuring your DUN connection profile

After you create a DUN connection profile, you can configure it for TCP/IP networking to enable your computer to connect to the Internet through the DSL modem. You use the Server Types dialog box to configure TCP/IP for the DUN profile. This dialog box includes a list of protocols for dial-up connections. You configure the TCP/IP protocol for Internet connections as follows:

1. **Double-click the My Computer icon on the desktop, and then double-click the Dial-Up Networking icon.**

   You see a new icon with the name you entered for your connection.

2. **Right-click the connection icon and choose Properties from the pop-up menu.**

   A connection profile dialog box appears with the name of your connection in the title bar.

3. **Click the Server Types tab (Figure 12-8).**

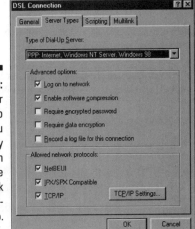

Figure 12-8:
The Server
Types tab
lets you
specify
information
about the
network
you're con-
necting to.

4. **In the Type of Dial-Up Server list, Windows 95 users should select PPP: Windows 95, Windows NT 3.5, Internet. Windows 98 users should select PPP: Internet, Windows NT Server, Windows 98.**

5. **In the Allowed network protocols area, click to add a check mark to the TCP/IP item. If necessary, click to remove any other check marks in that area.**

6. **Click the TCP/IP Settings button.**

   The TCP/IP Settings dialog box appears, as shown in Figure 12-9.

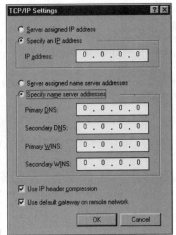

Figure 12-9:
Use the
TCP/IP
Settings
dialog box
to enter
the IP
addresses
of the
network
you're con-
necting to.

7. **Depending on the type of IP addressing used for your Internet access, do one of the following:**

   • If your ISP uses dynamic IP addressing, select the Server assigned IP address option at the top.

   • If your service provider has assigned you a specific IP address, select Specify an IP address. Enter the IP addresses assigned to your computer by the ISP, and then enter the IP address in the Primary DNS and Secondary DNS boxes.

8. **Click OK three times.**

Making the DUN connection

The easiest way to make a DUN connection is to double-click any TCP/IP application, such as the Internet Explorer or Netscape Communicator icon on your desktop. The browser window opens along with the Connect To dialog box. Click the Connect button. The connection is made and the home page appears in your Web browser. To disconnect, click the minimized Connect To dialog box on the taskbar and then click Disconnect.

You can access the DUN connection profile for your DSL connection also by opening the Dial-Up Networking folder in My Computer to display the Dial-Up Networking window (see Figure 12-10). Double-click the connection profile for your DSL connection. The Connect To dialog box appears (see Figure 12-11).

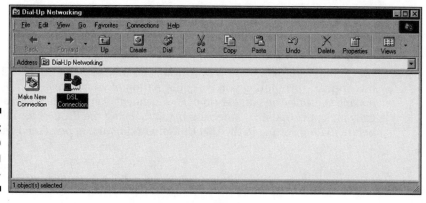

Figure 12-10:
The Dial-Up
Networking
window.

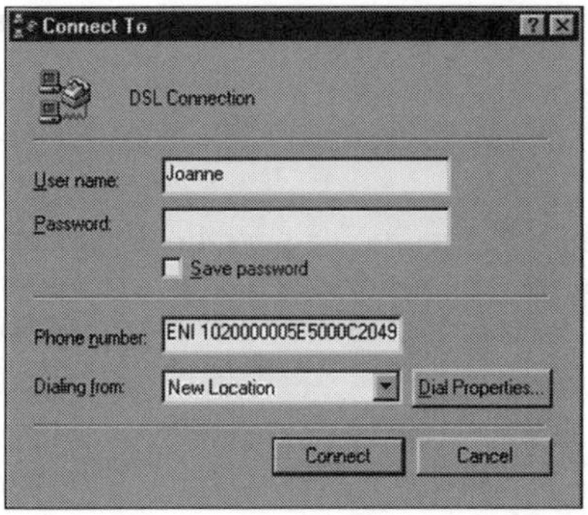

Figure 12-11:
The Connect
To dialog
box.

You need to enter your username and password in the appropriate fields. The Save password check box below the Password field, if checked, will tell Dial-Up Networking to remember the password so that you won't have to enter it on subsequent dial-up sessions. For a DSL connection, the number in the Phone number field is an encoded number or a virtual circuit identifier (VCI) number assigned by your ISP.

One of the big advantages of a DSL connection is that it is always on, so you will not be connecting and disconnecting as much as you did with a dial-up Internet connection. If you want to break a connection, however, display the Dial-Up Networking status box by clicking the DUN icon on the right side of the Windows 95/98 task bar. A Status dialog box appears; this dialog box varies depending on the DSL modem used. To terminate your session, click the Close button. The link terminates immediately.

Bypassing the Connect To dialog box

Every time you use a Dial-Up Networking connection, the Connect To dialog box appears, prompting you for confirmation of your connection before making the call. You can get rid of this Connect To dialog box and automatically make the dial-up connection by unchecking the Prompt for information before dialing setting in the Dial-Up Networking dialog box (see Figure 12-12).

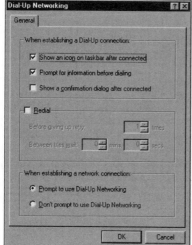

Figure 12-12:
The Dial-Up
Networking
dialog box.

Handy Windows 95/98 TCP/IP Utilities

Windows 95/98 has a collection of handy TCP/IP utilities that include winipcfg, telnet, ping, and tracert. These tools help you to manage and monitor your IP connection. This section explains how to use the TCP/IP utilities included with Windows 95/98.

Third-party utilities that offer more features are also available to help you. Chapter 8 goes into more detail on third-party tools to help you manage your DSL connection.

What's your IP status?

The winicfg program is a TCP/IP utility that lets you view information about your TCP/IP protocol and network adapter settings. If you're using DHCP on your LAN to receive an IP address from a host, the winipcfg program is a handy way to figure out the IP address that's been assigned to your PC and how long it's been assigned.

The winipcfg program is such a handy program to have accessible that you should have a shortcut to it on your Windows desktop. The easiest way to do this is to choose Start➪Find➪Files or Folders, enter **winipcfg.exe** in the Named field, select the drive containing the Windows directory, and then click Find Now. Drag the winipcfg file to the desktop. A shortcut to winipcfg now appears on your desktop.

The winipcfg program (winipcfg.exe) is located in your Windows folder. To run the winipcfg program, click StartÍRun. In the Open box, type **winipcfg.exe** and then click OK. The IP Configuration dialog box appears, as shown in Figure 12-13. Here, you can see the physical address, IP address, subnet mask, and default gateway settings for your primary TCP/IP network adapter card.

Figure 12-13:
The IP Config-
uration
dialog box.

Click the More Info button to display an expanded IP Configuration dialog box, as shown in Figure 12-14. This dialog box displays such information as your computer's host name, the address of the DHCP server (if you're using one), and DNS server IP addresses.

Figure 12-14:
The
expanded IP
Config-
uration
dialog box.

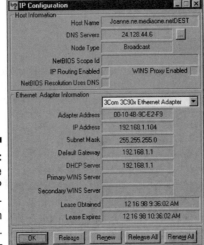

Information in the expanded IP Configuration dialog box is broken down into two groups: Host Information and Ethernet Adapter Information. Also included is information on a variety of IP and DNS settings and network adapter information (such as the hardware address). The Release and Renew buttons at the bottom of the dialog box enable you to release and renew the assigned IP address, respectively.

You can print the information in the IP Configuration dialog box by copying the data to Word or NotePad. To do this, click the Winipcfg control menu, which is the small icon in the window title bar (upper-left corner) of the IP Configuration window. Choose Copy or press Ctrl+C. The contents of the Winipcfg window are copied into the Windows Clipboard, enabling you to paste the information into another application. After the information is in the application (such as Word or Notepad), you can manipulate the text or print it.

Reach out and telnet someone

The telnet program is a terminal emulator program that enables you to connect to and log on to a remote host, where you can perform tasks such as running programs on the remote host. It's a handy tool for managing a DSL router or modem remotely because you can typically access the device using telnet from any workstation on your LAN.

You can create a shortcut to the telnet program on your desktop in the same way you create a shortcut for the winipcfg program. Check out the section called "What's your IP status?" earlier in this chapter for more info on creating a shortcut.

The telnet program (telnet.exe) is in the Windows folder. To check out the program, follow these steps:

1. **Open the telnet program by navigating to your Windows directory, and then double-click the telnet.exe file.**

   The telnet window appears.

2. **Choose Connect⇨Remote System.**

   The Connect dialog box appears, as shown in Figure 12-15.

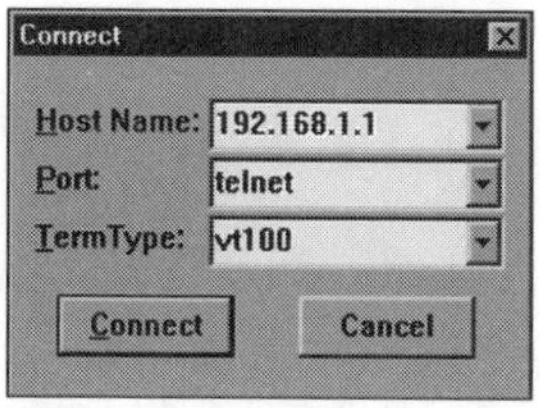

Figure 12-15:
The Connect
dialog box.

Connect	☒
Host Name:	192.168.1.1
Port:	telnet
TermType:	vt100
Connect	Cancel

3. **In the Host Name field, type the IP address or domain name of your DSL router.**

For example, if the IP address of your DSL router is 199.232.255.113, type that in the field.

4. **Click Connect.**

Connecting to a Netopia 9100 router, for example, displays the screen shown in Figure 12-16.

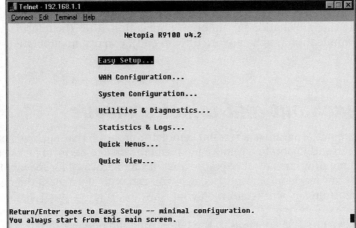

Figure 12-16: The telnet window displaying a menu for the Netopia R9100 Router.

5. **To disconnect from the telnet session, choose Connect⇨Disconnect.**

After you've made a connection using telnet, the IP address or host name appears in the Connect menu, so you can quickly access the device again without reentering the information in the Connect dialog box.

Ping, ping, ping

The ping program is a handy TCP/IP diagnostic utility that works like TCP/IP sonar: You send a packet to a remote host, and if the host is functioning properly, it bounces the packet back to you. Ping prints the results of each packet transmission on your screen. By default, Ping sends four packets, but you can use it to transmit any number of packets or transmit continuously until you terminate the command.

You can access ping from the MS-DOS Prompt window (Start⇨Programs⇨ MS-DOS Prompt). At the DOS prompt, enter the command in the following format:

```
ping IPaddress
```

where *IPaddress* is the IP address of the remote host you want to check. For example, if you type the following:

```
ping 199.232.255.113
```

and press Enter, you ping the host machine with the IP address of 199.232.255.113. The host machine can be on your local network or on the Internet. Likewise, you can ping the host name instead of the IP address for sites for which you don't know the IP address. For example, entering the following:

```
ping www.angell.com
```

returns the same ping information that you would get using the IP address. Figure 12-17 shows a sample ping result as it appears in the MS-DOS Prompt window.

Figure 12-17: The MS-DOS Prompt window with ping results.

If the packets don't come back after you execute the ping command, either the host is not available or something is wrong with the connection.

> The ping program is the single most useful program for troubleshooting the status of any TCP/IP connection. The ping command includes a number of parameters, which you can view by typing **ping** and pressing Enter.

Tracing your routes

The tracert (trace route) command is an extension to ping that lets you trace the route your data is taking to get to a specific site. The tracert command shows you how many routers are between your computer and the specified server, and displays the names of the routers used to route your data. The number of routers used between a client and a server on the Internet is referred to as the number of *hops*.

Type **tracert** followed by the IP addresses or host name of the site, and the screen displays the number of hops and the connection times. For example, when I typed **tracert www.angell.com**, the information from my location to the Web server for www.angell.com was displayed, as shown in Figure 12-18.

Figure 12-18:
The tracert command allows you to see how many hops your data must take to get to a host.

Chapter 13

Windows NT Meets DSL

● ●

In This Chapter

▶ Connecting DSL to Windows NT

▶ Finding out how to set up TCP/IP in Windows NT

▶ Discovering how to configure NICs for TCP/IP

▶ Installing and configuring Windows NT Remote Access Service

▶ Using Dial-Up Networking for DSL connections

▶ Working with some handy Windows NT TCP/IP utilities

● ●

To use a computer running Windows NT 4.0 (Workstation or Server) to access the Internet via DSL, you need to install and configure the Microsoft TCP/IP stack. If you're using a DSL bridge or router to connect to the Internet, you need to configure the TCP/IP setting for your network interface card. If you're using a DSL modem card or USB modem with Windows 95/98, you'll also have to install the Microsoft TCP/IP stack, and you may also need to install and configure Windows NT's Remote Access Service and Dial-Up Networking. This chapter explains the fundamentals of configuring Windows NT for connecting to your DSL CPE.

If you're using Windows 95 or 98 for your DSL connection rather than Windows NT, read no further. Instead, check out Chapter 12.

Windows NT and DSL Connections

As explained in Chapter 5, you can use different types of DSL CPE as part of your Internet service. These different types of DSL CPE interface differently to Windows NT. If you use a DSL modem adapter card (PCI card) or an external USB (Universal Serial Bus) modem, you use Windows NT's Dial-Up Networking to make the Internet connection through DSL. If you use an external DSL bridge or router, Windows NT uses the Ethernet interface through a NIC (network interface card). Using an Ethernet solution means you configure the NIC card for TCP/IP.

You can connect a Windows NT computer to DSL in three ways:

- A DSL modem adapter card (PCI card)
- An external USB modem
- The Ethernet interface through a NIC

With a DSL bridge or router, you install and configure the Microsoft TCP/IP stack for every Windows NT computer on the LAN. You configure the TCP/IP properties of the NIC to enable your DSL connection.

If you're using a PCI or a USB DSL modem that uses DUN, it fools Windows NT into thinking it is doing a dial-up connection, although DSL isn't a dial-up service. This means you'll still be using DUN in Windows NT to work with either form of DSL CPE. This is a function of the DSL connection being handled as PPP (Point-to-Point Protocol), which is the standard protocol for dial-up connections to the Internet. A DSL modem can connect to your PC using PPP (Point-to-Point Protocol) in two ways: PPP over ATM and PPP over Ethernet (PPPoE). See Chapter 3 for more information on these DSL PPP protocols.

When you install many PCI DSL modem cards or USB modems, they're treated as NICs in Windows NT. The software driver for the DSL modem is installed using the same Network properties dialog box used to install NIC cards. Other PCI DSL modem cards and USB modems are installed as modem drivers in the Modem properties dialog box.

Check the power under the hood

Windows NT requires certain minimum hardware capabilities to operate. Connecting to a high-speed DSL connection can also demand more from your PC. Higher bandwidth requires more PC power on several fronts, most notably the CPU, RAM, and video card. Following are some guidelines for the optimal configuration of a Windows NT computer:

- Use a Pentium or Pentium II class processor. More is always better when it comes to processor power. Most of today's PCs meet these requirements.

- Use an accelerated PCI bus video card that supports higher resolutions. Accelerated display cards make a difference, and the more onboard memory on the display card, the better. Larger monitors and higher resolution give you more real estate for multiple applications or larger viewing areas. Most new video cards easily surpass these minimum standards.

- At least 32MB of RAM, with 64MB the norm or 128MB if you plan to use Windows NT Server for running services, such as for a Web server.

Setting Up TCP/IP in Windows NT

Setting up Windows NT for a DSL connection involves installing and configuring the Microsoft TCP/IP stack, which provides the TCP/IP networking
protocol for connecting to the Internet. When you install the TCP/IP stack,
you also bind the TCP/IP protocol to your network adapter card and Dial-Up
Networking (if it's installed), which is part of Windows NT's Remote Access
Service. *Binding* is a process that defines the relationship between a network
adapter card and network protocols used with that network card. When you
bind the TCP/IP protocol to a network adapter, you allow TCP/IP traffic to
pass through your Ethernet network to the DSL modem or router. Likewise,
binding TCP/IP to the Dial-Up Networking adapter enables you to use TCP/IP
for DSL connections using Dial-Up Networking.

The first step in configuring Windows NT is to install the Microsoft TCP/IP
stack, if it's not already installed. If you've been connecting to the Internet,
Microsoft TCP/IP is probably running.

You can check to see whether the TCP/IP stack is installed by right-clicking
Network Neighborhood, and choosing the Protocols tab (see Figure 13-1). If
you see TCP/IP Protocol in the Network Protocols list, you already have the
Microsoft TCP/IP stack installed. You can skip the following steps but you still
need to make sure that TCP/IP is bound to the NIC, which is explained later in
this chapter.

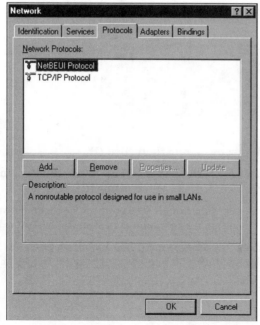

Figure 13-1:
The
Protocols
tab in the
Network
properties
dialog box
shows what
networking
protocols
you have
installed.

If you don't have TCP/IP installed, you need to install the Microsoft TCP/IP stack. Installing the Microsoft TCP/IP stack binds TCP/IP to any installed NIC or DSL modem driver in the Network properties dialog box. Before installing the Microsoft TCP/IP stack, have the original Windows NT distribution media handy, in case Windows prompts you for it.

To install the Microsoft TCP/IP stack, do the following:

1. **Right-click the Network Neighborhood icon on the desktop and choose Properties from the pop-up menu.**

   Alternatively, you can double-click the Network icon in the Control Panel. The Network properties dialog box appears.

2. **Click the Protocols tab to display the Protocols properties.**

3. **Click Add.**

   The Select Network Protocol dialog box appears, as shown in Figure 13-2.

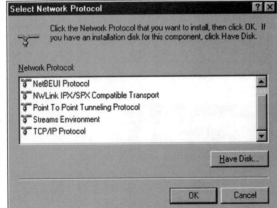

Figure 13-2:
The Select
Network
Protocol
dialog box.

4. **Select the TCP/IP Protocol, and then click OK.**

5. **When Windows NT asks whether you want to use the Dynamic Host Configuration Protocol (DHCP) when using TCP/IP, click No.**

   If you add DHCP server functionality to your LAN, you can always change the TCP/IP settings, as explained later in this chapter.

6. **If you have Remote Access Service (RAS) installed, click Yes when Windows NT asks whether you want to configure it to use TCP/IP.**

7. **When Windows NT displays the Windows NT Setup dialog box to prompt you for the location of the distribution files, enter the drive and directory where the Windows NT 4.0 distribution files are located.**

8. **Click Continue.**

   After the files are copied, Windows NT displays the Network dialog box with the TCP/IP Protocol visible.

9. **Click Close to complete the installation.**

   After the bindings have been configured, Windows NT displays the Network Settings Change dialog box to notify you that you must restart your computer before the changes will take effect.

10. **Click Yes to restart the computer.**

Configuring NICs for TCP/IP in Windows NT

When you bind the TCP/IP protocol to a network adapter, you allow TCP/IP traffic to pass through your Ethernet network to the DSL bridge or router. You need to configure the Microsoft TCP/IP stack bound to the NIC for the IP address configuration you're using on your LAN.

You can configure your TCP/IP stack in two ways:

✔ If you're using a DSL router as a DHCP (Dynamic Host Configuration Protocol) server to dynamically assign IP addresses, configure your Windows NT computers as DHCP clients, and don't add any specific IP address information to the TCP/IP properties.

✔ If you're using static IP addresses assigned to specific computers, manually configure the TCP/IP properties of each Windows NT computer with IP address and DNS information.

Windows NT allows only one TCP/IP stack configuration profile for LAN-to-Internet (or other network) connections. This means if you have more than one ISP connection available, you must change the settings in the TCP/IP Properties dialog box each time and then reboot the system. A shareware program called NetSwitcher (www.netswitcher.com) allows you to create multiple TCP/IP network connections and switch between them as needed. For example, if you're using a laptop and travel between different offices, each with its own LAN-based Internet access service, you can use NetSwitcher to create profiles for all of them and switch to the one you need when you're at a specific office. NetSwitcher also makes it easy to switch back and forth between a dedicated DSL connection and a dial-up connection when on the road.

Configuring Windows NT for dynamic IP addressing

Using a DSL router that acts as a DHCP (Dynamic Host Configuration Protocol) server enables dynamic IP addressing. The DHCP server assigns an IP address from an available pool of IP addresses for use only during the current connection. The DHCP server functionality of a router allows IP addresses, subnet masks, and default gateway addresses to be assigned to computers on your LAN. You can use DHCP with registered IP addresses or private IP addresses along with the NAT (network address translation) feature incorporated in most DSL routers. Chapter 4 explains IP addressing and DHCP in more detail.

Each Windows NT computer on your LAN with the Microsoft TCP/IP stack installed is a DHCP client out of the box. If you've installed TCP/IP for the first time, you don't need to change any default settings to support DHCP. However, if you're unsure of your TCP/IP configuration, you can double-check to make sure the default TCP/IP settings are in place by performing the following steps:

1. **Right-click the Network Neighborhood icon and choose Properties from the pop-up menu.**

   The Network properties dialog box appears.

2. **Click the Protocols tab.**

3. **Select the TCP/IP option, and then click the Properties button.**

4. **Click the IP Address tab. If necessary, select the adapter card if you're using more than one card.**

5. **Select the Obtain an IP Address from a DHCP Server option.**

6. **Click OK twice to exit the Network applet.**

7. **Restart Windows NT.**

After you configure your Windows clients for DHCP, whenever you boot the Windows NT computer, it will get the IP address information it needs from the DHCP server.

Configuring Windows NT for static IP addressing

If you plan on using registered IP addresses that you want to assign to specific computers to, for example, run a Web server, you need to configure the Windows NT computer with static IP address and DNS information. You can also use private (unregistered IP addresses) for your static IP addressing.

Before you configure your Windows NT computers for static IP addressing, you need to get the following information from your Internet service provider (ISP):

✔ An IP address for each Windows NT computer. In most cases, these IP addresses will be registered IP addresses.

✔ A subnet mask IP address for your network.

✔ The IP address assigned to your DSL bridge or router, which is the gateway IP address.

✔ A host and domain name for the registered IP addresses.

✔ DNS server IP addresses, which are typically IP addresses supplied by the ISP.

To set up a Windows NT (Server or Workstation) for static IP addressing, do the following:

1. **In the Control Panel, double-click the Network icon.**

   The Network dialog box appears.

2. **Click the Protocols tab, select the TCP/IP Protocol option, and then click Properties.**

   The Microsoft TCP/IP Properties dialog box appears (see Figure 13-3). If you have only one network adapter installed in your computer, Windows NT displays the adapter name in the Adapter list.

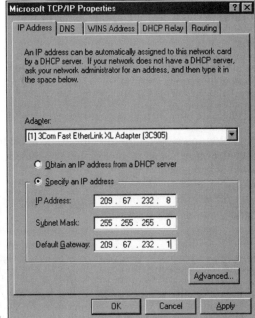

Figure 13-3:
The
Microsoft
TCP/IP
Properties
dialog box.

3. If you have more than one adapter installed, use the drop-down list to select the adapter for which you want to configure TCP/IP properties.

4. Click the IP Address tab.

5. Select the Specify an IP address option.

6. Type the IP address and subnet mask IP address in their respective boxes.

7. In the Default gateway field, type the IP address of your router or bridge.

8. Click the DNS tab.

9. In the Host Name box, type the host name of the client. In the Domain box, type the domain name.

10. Type the IP address of the ISP's DNS server, and then click the Add button. Repeat for each additional DNS server address supplied by your ISP.

11. Click OK twice to exit the Microsoft TCP/IP Properties and Network dialog boxes.

12. If prompted, insert your distribution disk or CD-ROM.

13. When prompted, click Yes to restart Windows NT.

Beyond basic TCP/IP configuration

Windows NT has more sophisticated TCP/IP capabilities than those in Windows 95/98. Windows NT's advanced IP addressing includes the following:

✔ Adding multiple IP addresses to a single NIC

✔ Choosing from multiple gateways

✔ Enabling a Microsoft virtual private network (VPN) solution

✔ Specifying TCP/IP security parameters

You can configure Windows NT's advanced IP addressing features by doing the following:

1. **Right-click the Network Neighborhood icon and choose Properties.**

2. **Click the Protocols tab, select TCP/IP Protocol, and click the Properties button.**

3. **Click the Advanced button.**

   The Advanced IP Addressing dialog box appears, as shown in Figure 13-4.

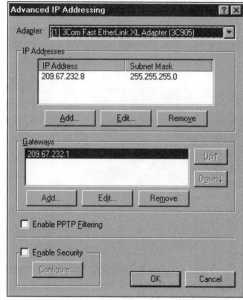

Figure 13-4:
The
Advanced
IP
Addressing
dialog box.

4. **Use the Advanced IP Addressing dialog box to do any of the following:**

   - In the IP Addresses group, use the Add, Edit, and Remove buttons to modify the IP addresses assigned to the selected network adapter. You can specify as many as 16 IP addresses per network adapter. You can use multiple IP addresses to make a single computer appear to the network as multiple virtual computers. This is useful if you want to run multiple independent Web servers on a single computer, with each Web server discretely addressable.

   - In the Gateways group, use the Add, Edit, and Remove buttons to modify the gateway's configuration. Gateways are searched in the order they are listed in the Gateways list (from top to bottom). If more than one gateway is listed, you can use the Up and Down buttons to modify the search order listing.

   - Click to put a check mark in the Enable PPTP Filtering option, and you activate Microsoft's virtual private network (VPN) solution, called Point-to-Point Tunneling Protocol (PPTP). See Chapter 8 for more on VPNs.

   - Click to put a check mark in the Enable Security option, and you can specify TCP/IP security parameters by clicking the Configure button to display the TCP/IP Security dialog box (see Figure 13-5). You can specify TCP Ports, UDP ports, and IP protocols on your computer that are available to the network. By default, all ports and protocols are available. You might, for example, set these parameters to allow other computers to access a Web server running on your computer while prohibiting them from accessing other TCP/IP services.

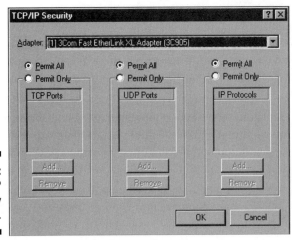

Figure 13-5:
The TCP/IP
Security
dialog box.

5. **After you have finished making changes in the Advanced IP Addressing, click OK to return to the Microsoft TCP/IP Properties dialog box.**

Working the DUN Way

The Windows NT Remote Access Service (RAS) is the basis for any dial-up connection. RAS includes a Dial-Up Networking (DUN) facility similar to the one used by Windows 95/98, although it does have differences in implementation. The NT version of DUN is fused to RAS on the client side. Dial-Up Networking replaces the dial-out component of RAS, but during the installation and configuration of DUN, you'll see dialog boxes and references connected to RAS.

If you're using a DSL modem card or a USB modem that requires you to use Windows Dial-Up Networking, you need to install RAS. After you install and configure RAS for DUN, you then create a connection profile. The DUN connection profile provides Windows NT with the information it needs to make the DSL connection. Some DSL modem vendors create the profile automatically as part of the installation, which uses information added to the software by the DSL Internet service provider.

The creation of a DUN connection profile mimics the creation of a DUN profile for an analog or ISDN modem — with one major difference. The entry for the telephone number is a virtual circuit (or channel) identifier (VCI) number. This 16-bit number is provided by the DSL ISP and identifies the channel your DSL service will use to make the connection.

Installing and configuring RAS for DUN

Windows NT Remote Access Service (RAS) is not installed automatically when you install Windows NT. With your DSL modem installed, you need to install RAS and configure it for Dial-Up Networking (DUN). You install RAS from your original Windows NT distribution media (CD-ROM) or from the hard drive, if you copied the original files there.

To install and configure RAS, do the following:

1. **Open the My Computer folder on the desktop, and then double-click the Dial-Up Networking icon.**

   The Dial-Up Networking dialog box appears, asking whether you want to install the software.

2. **Click Install.**

   The Windows NT Setup dialog box appears.

3. **Enter the location of your Windows NT original media.**

   This is usually D:\i386, where D is the drive letter of your CD-ROM and i386 is the directory for Windows NT files for Intel platforms. If you copied the files to your hard drive, use the appropriate directory path instead of the path to your CD-ROM.

4. **Click Continue.**

   The necessary files are transferred to your hard drive. Then the Add RAS Device dialog box automatically appears, and any ports that contain devices that can be used for DUN are listed in the RAS Capable Devices dialog box.

5. **In the RAS Capable Devices list, select the DSL modem, and then click OK.**

   The Remote Access Setup dialog box appears.

6. **Select the Port/Device setting, and then click the Configure button.**

   The Configure Port Usage dialog box appears.

7. **Click to add a check mark to the Dial out only option, and then click OK.**

8. **Click Close to exit the Network dialog box.**

9. **Restart Windows.**

Creating a Phonebook entry

You can create a Dial-Up Networking profile using the Dial-Up Networking icon or the Internet Connection Wizard. A DUN profile includes all the specific IP address information you need to make a connection to the ISP.

You create a DUN profile as a Phonebook entry using the Dial-Up Networking icon:

1. **Double-click the My Computer icon on the desktop, and then double-click the Dial Up Networking icon.**

   The first time you open the Phonebook, a message box appears telling you that the phonebook is empty.

2. **Click OK.**

   The New Phonebook Entry Wizard appears.

3. **Type a name to identify your phonebook entry, and then click Next.**

4. **Select the I am calling the Internet option, and then click Next.**

5. **In the Phone Number field, enter the virtual circuit identifier number, and then click Next.**

6. **Click Finish.**

   The Dial-Up Networking dialog box appears, as shown in Figure 13-6. After you create your first Phonebook entry, the next time you double-click the Dial-Up Networking icon in the My Computer folder, the Dial-Up Networking dialog box appears instead of the New Phonebook Entry Wizard.

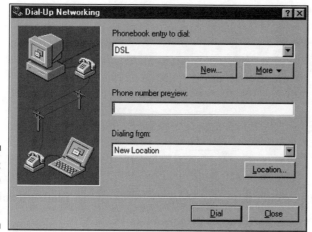

Figure 13-6: The Dial-Up Networking dialog box.

Configuring your Phonebook entry for TCP/IP

After you create a Phonebook entry, you configure it for TCP/IP networking so that you can use it to connect to the Internet through the DSL modem and DSL connection. You use the Server properties page to configure TCP/IP for the Phonebook entry. This page includes a list of protocols for dial-up connections. You configure the TCP/IP protocol for Internet connections, as follows:

1. **In the Dial-Up Networking dialog box, click the More button.**

   A menu appears.

2. **From the menu, choose Edit Entry and Modem Properties.**

   The Edit Phonebook Entry dialog box appears.

3. **Click the Server tab to display the server properties (see Figure 13-7).**

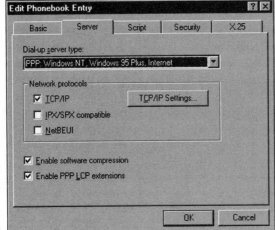

Figure 13-7:
The Server
properties
page in the
Edit
Phonebook
Entry dialog
box.

4. **In the list, select PPP: Windows NT, Windows 95 Plus, Internet.**

5. **In the Network protocols group, select TCP/IP.**

   Make sure the other two options in the group are not selected.

6. **If you're using DHCP, click OK to return to the Dial-Up Networking dialog box.**

7. **If you're using a static IP address, follow these instructions:**

   a. **Click the TCP/IP Settings button.**

      The PPP TCP/IP Settings dialog box appears, as shown in Figure 13-8.

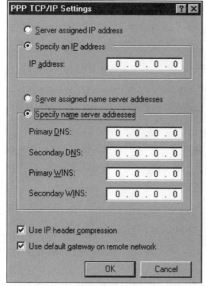

Figure 13-8:
The PPP
TCP/IP
Settings
dialog box.

b. **Select the Specify an IP address option, and enter the IP address for the computer.**

c. **Select the Specify name server addresses option, and enter the primary and secondary DNS server IP addresses.**

d. **Click OK twice to return to the Dial-Up Networking dialog box.**

Keeping up appearances

Because DUN is designed primarily for dial-up sequences used by modems, it displays a number of dialog message boxes to keep you apprised of the status of your dial-up connection. In most cases, you want to shut off these displays because the connection is instant and you don't want to be prompted for a response each time you make a connection.

In the Dial-Up Networking dialog box, clicking the More button and choosing User preferences will display the User Preferences dialog box. For DSL dial-up connections, the only relevant item in the User Preferences dialog box is the Appearance tab, shown in Figure 13-9.

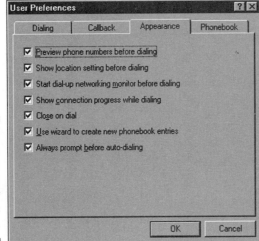

Figure 13-9:
The
Appearance
tab in the
User
Preferences
dialog box.

The Appearance tab allows you to set several options that control the appearance and function of DUN. Table 13-1 lists these settings and how you should configure them for working with a DSL connection.

Table 13-1   **Settings in the Appearance Page of the User Preferences Dialog Box**

Option	Recommended Setting	What It Does
Preview phone numbers before dialing	Unchecked	Displays and allows you to change the telephone number to be dialed before dialing occurs
Show location setting before dialing	Unchecked	Displays and allows you to change the location before dialing occurs
Start dial-up networking monitor before dialing	Unchecked	Activates DUN monitor automatically each time a DUN session is started
Show connection progress while dialing	Uncheck	Displays the progress of each DUN call step-by-step as the connection occurs and the session is established

(continued)

Option	Recommended Setting	What It Does
Close on dial	Checked	Closes the Dial-Up Networking dialog box after dialing commences
Use Wizard to create phonebook entries	Checked/Unchecked	Uses the Dial-Up new Networking Wizard to create new Phonebook entries; unchecking this setting lets you create new DUN entries using the standard dialog box instead of the Wizard
Always prompt before auto-dialing	Unchecked	Prompts you before auto-dialing a DUN Phonebook entry

Making the connection

After you've created your Dial-Up Networking Phonebook entry, you're ready to make your connection. You can initiate a DUN connection in two ways. One way is to simply double-click the Web browser icon. The browser window opens along with the Connect To dialog box. Click the Connect button. The connection is made, and the home page appears in your Web browser. To disconnect, click the minimized Dial-up Connection dialog box on the taskbar and then click Disconnect.

Another way to make the Dial-Up Networking connection is shown in the following steps:

1. **In the My Computer folder, double-click the Dial-Up Networking icon.**

2. **In the Dial-Up Networking dialog box, click the Dial button.**

   The Dial-up Connection dialog box appears, as shown in Figure 13-10.

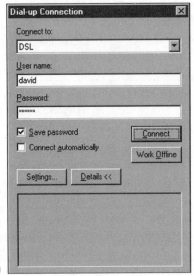

Figure 13-10:
The Dial-up
Connection
dialog box.

3. **Type your username and password for logging on to the Internet, if necessary.**

4. **Check the Save Password box to save the password so you don't have to enter it each time.**

You can add or change your username and password using the Internet Properties dialog box.

5. **Click the Settings button.**

The Dial-Up Settings dialog box appears.

6. **Click the Connect button to make your connection.**

You can monitor connections using the Dial-Up Networking Monitor, which is present on the status bar for the duration of the connection. To invoke the Monitor, double-click its icon. The Dial-Up Networking Monitor window appears. To end your connection, click Disconnect.

Handy Tools of the IP Trade

Windows NT has a collection of handy TCP/IP utilities that include ipconfig, telnet, ping, and trace route. These tools help you to manage and monitor your IP connection. This section explains how to use the TCP/IP utilities included with Windows NT.

Third-party utilities that offer more features are also available to help you. Chapter 8 goes into more detail on third-party tools to help you manage your DSL connection.

What's your IP status?

The ipconfig program is available in Windows NT as a command-line program, although the Windows NT Resource Kit includes a graphics version similar to one in Windows 95/98. This TCP/IP utility lets you view information about your TCP/IP protocol and network adapter settings. If you're using DHCP on your LAN to receive an IP address from a host, the ipconfig program is a handy way to know the IP address that has been assigned to your PC and how long it's been assigned.

To use the ipconfig program, choose Start⇨Programs⇨Command Prompt. The Command Prompt window appears. Typing **ipconfig** displays just the IP address, subnet mask, and default gateway information, as shown in Figure 13-11. Typing **ipconfig/all** lists all the IP address attributes. You can type **ipconfig/?** to display a list of all the commands you can use with ipconfig.

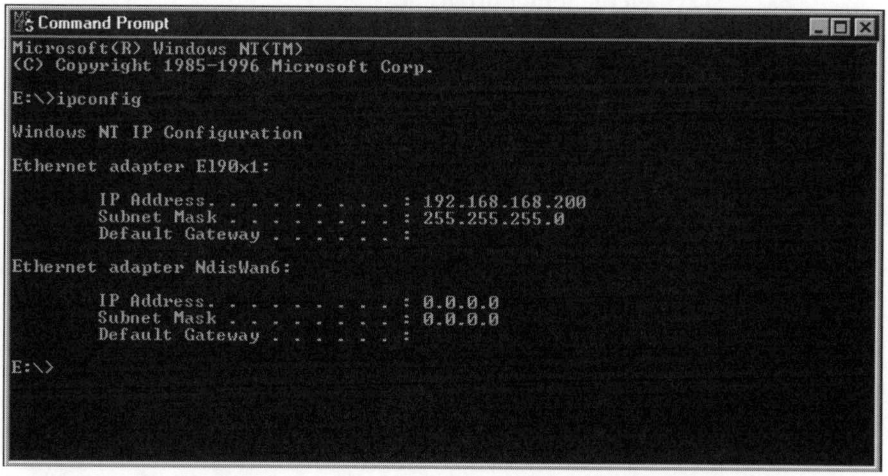

Figure 13-11:
The results of entering the ipconfig command in the Command Prompt window.

```
Command Prompt                                                    _ □ ×
Microsoft(R) Windows NT(TM)
(C) Copyright 1985-1996 Microsoft Corp.

E:\>ipconfig

Windows NT IP Configuration

Ethernet adapter E190x1:

        IP Address. . . . . . . . . : 192.168.168.200
        Subnet Mask . . . . . . . . : 255.255.255.0
        Default Gateway . . . . . . :
Ethernet adapter NdisWan6:

        IP Address. . . . . . . . . : 0.0.0.0
        Subnet Mask . . . . . . . . : 0.0.0.0
        Default Gateway . . . . . . :

E:\>
```

Reach out and telnet someone

The telnet program is a terminal emulator program that enables you to connect to and log on to a remote host, where you can perform tasks such as running programs on the remote host. It's a handy tool for managing a DSL router or modem remotely because you can typically access the device using telnet from any workstation on your LAN.

The telnet program (telnet.exe) is located in the \Winnt\system32 folder. The easiest way to locate the telnet program is as follows:

1. **Choose Start⇨Find⇨Files or Folders.**

   The Find All Files dialog box appears.

2. **In the Named field, type** telnet.exe.

3. **In the Look in field, select the drive that contains Windows NT.**

4. **Click the Find Now button.**

   The telnet.exe file appears in the list.

   To create a handy shortcut, drag the icon to the desktop.

5. **Double-click the telnet program.**

   The telnet window appears.

To use telnet, the computer you're using to connect to the DSL device must be part of the same IP network. Here is how to use telnet to access a DSL device:

1. **Open the telnet program.**

   If you followed the last set of steps, the program is already open.

2. **Choose Connect⇨Remote System.**

   The Connect dialog box appears, as shown in Figure 13-12.

Figure 13-12:
The Connect
dialog box.

Connect	✕
Host Name:	192.168.1.1 ▾
Port:	telnet ▾
TermType:	vt100 ▾
Connect	Cancel

3. **In the Host Name field, type the IP address or domain name of your DSL router.**

   For example, if the IP address of your DSL router is 199.232.255.113, type that in the field.

4. **Click Connect.**

   Connecting to a Netopia 9100 router, for example, displays the screen shown in Figure 13-13.

Figure 13-13:
The telnet
window dis-
playing a
menu for the
Netopia
R9100
Router.

5. **To disconnect from the telnet session, choose Connect⇨Disconnect.**

   After you've made a connection using telnet, the IP address or host name appears in the Connect menu, so you can quickly access the device again without reentering the information in the Connect dialog box.

Ping, ping, ping

The ping program is a handy TCP/IP diagnostic utility that works like TCP/IP sonar: You send a packet to a remote host, and if the host is functioning properly, it bounces the packet back to you. Ping prints the results of each packet transmission on your screen. By default, Ping sends four packets, but you can use it to transmit any number of packets or transmit continuously until you terminate the command.

You can access ping from the Command Prompt window (Start⇨Programs⇨Command Prompt). At the DOS prompt, enter the command in the following format:

```
ping IPaddress
```

where *IPaddress* is the IP address of the remote host you want to check. For example, if you type the following:

```
ping 199.232.255.113
```

and press Enter, you ping the host machine with the IP address of 199.232.255.113. The host machine can be on your local network or on the Internet. Likewise, you can ping the host name instead of the IP address for sites for which you don't know the IP address. For example, entering the following:

```
ping www.angell.com
```

returns the same ping information that you would get using the IP address. Figure 13-14 shows a sample ping result as it appears in the Command Prompt window.

```
Command Prompt                                              _ □ ×
Microsoft(R) Windows NT(TM)
(C) Copyright 1985-1996 Microsoft Corp.

D:\>ping 209.67.232.1

Pinging 209.67.232.1 with 32 bytes of data:

Reply from 209.67.232.1: bytes=32 time=50ms TTL=46
Reply from 209.67.232.1: bytes=32 time=50ms TTL=48
Reply from 209.67.232.1: bytes=32 time=50ms TTL=48
Reply from 209.67.232.1: bytes=32 time=50ms TTL=48

D:\>
```

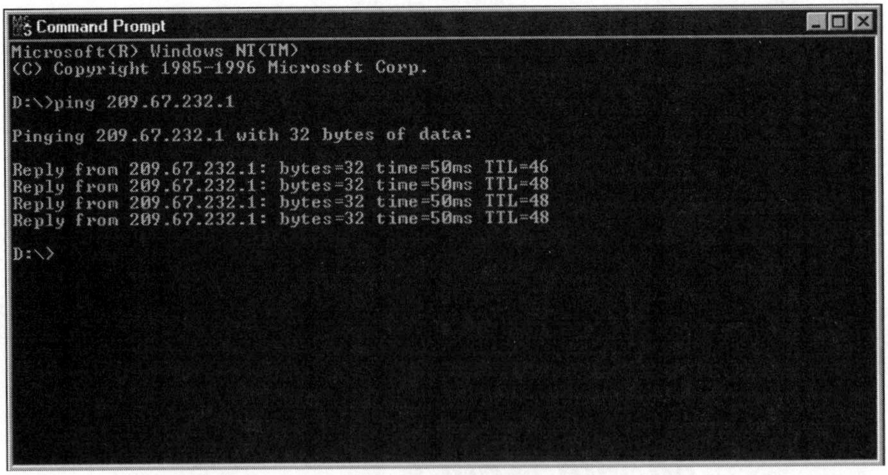

Figure 13-14:
The
Command
Prompt
window with
ping results.

If the packets don't come back after you execute the ping command, either the host is not available or something is wrong with the connection.

The ping program is the single most useful program for troubleshooting the status of any TCP/IP connection. The ping command includes a number of parameters, which you can view by typing **ping** and pressing Enter.

Tracing your routes

The tracert (trace route) command is an extension to ping that lets you trace the route your data is taking to get to a specific site. The tracert command shows you how many routers are between your computer and the specified server, and displays the names of the routers used to route your data. The number of routers used between a client and a server on the Internet is referred to as the number of *hops*.

Type **tracert** followed by the IP addresses or DNS name of the site, and the screen displays the number of hops and the connection times. For example, when I typed tracert www.angell.com, the information from my location to the Web server for www.angell.com was displayed, as shown in Figure 13-15.

Figure 13-15: The tracert command allows you to see how many hops your data must take to get to a host.

```
Command Prompt                                                    _ □ ✕
D:\>tracert 209.67.232.1

Tracing route to 209.67.232.1 over a maximum of 30 hops

  1    <10 ms    <10 ms    <10 ms  216.112.41.241
  2     20 ms     20 ms     20 ms  206.83.85.50
  3     10 ms     10 ms     10 ms  206.83.85.49
  4     40 ms     40 ms     41 ms  rt001a0601.chi-il.concentric.net [206.83.91.145]
  5     40 ms     40 ms     40 ms  rt001a0601.chi-il.concentric.net [206.83.91.145]
  6     90 ms     90 ms     90 ms  207.88.0.177
  7     90 ms    100 ms    101 ms  us-ca-pa-core1-a1-0d5.rtr.concentric.net [207.88
.0.14]
  8    100 ms    100 ms    110 ms  paix.exodus.net [198.32.176.15]
  9    110 ms    110 ms    110 ms  bbr02-p0-2.sntc03.exodus.net [209.185.249.181]
 10      *        110 ms    110 ms  dcr02-p12000.sntc03.exodus.net [209.185.9.254]
 11    111 ms    140 ms    110 ms  dcr02-p12000.sntc03.exodus.net [209.185.9.254]
 12    131 ms    120 ms    180 ms  dcr01-s02100.rvd101.exodus.net [209.185.9.162]
 13    121 ms    130 ms    120 ms  dcr01-h01000.vlhm01.exodus.net [209.185.9.166]
 14    141 ms    130 ms    130 ms  209.67.232.1

Trace complete.

D:\>
```

Part IV
The Part of Tens

The 5th Wave   By Rich Tennant

"THE PHONE COMPANY BLAMES THE MANUFACTURER, WHO SAYS IT'S THE SOFTWARE COMPANY'S FAULT, WHO BLAMES IT ON OUR MOON BEING IN VENUS WITH SCORPIO RISING."

In this part . . .

As its name implies, the "Part of Tens" imparts tens upon tens of valuable resource nuggets to help you on your way toward DSL enlightenment. Much of this information enhances and supports topics covered in previous chapters. You'll find helpful information on why you should consider DSL service, tips on how to shop for the best DSL service, and the questions to ask about DSL service.

Chapter 14

Dave's Top Ten Reasons for Getting DSL Service

A fatter pipeline to the Internet is only part of the DSL connection story. High-speed, always-on DSL service changes your Internet experience, enabling businesses, teleworkers, and consumers to get more from the Internet. Here are the top ten reasons for considering DSL service.

Cut Your Web Surfing Times

Surfing the Web through DSL brings it to life in all its multimedia glory. Faster Web access is the number one reason for DSL service. You can do more in less time, which can make you more productive.

Simply double-clicking the Microsoft Internet Explorer or Netscape Communicator icon on your Windows desktop delivers the Web to your desktop. No more squealing modem connections that seem to take forever. No waiting for Web page downloads. In fact, your connection may be so fast that you'll begin to see how slow different Web servers are in delivering information to you. You might find yourself waiting for a Web server to deliver data instead of waiting for your connection to do so.

Move Big Files Quickly

Increasingly, the Internet is being used as the software distribution method of choice for entire programs and software updates. Your DSL connection can become the main tool for getting and managing your software. Desktop publishers, multimedia developers, software publishers, and a host of others can deliver digital content economically through DSL connections either as a download or by providing file servers.

The more reliable digital nature of DSL service combined with raw speed makes the movement of these larger files a hassle-free experience. For example, downloading a 72MB file takes 25 minutes at 56 Kbps, 10 minutes at 128 Kbps, and only 48 seconds at 1.5 Mbps.

Run Your Own Web Server

A DSL connection combined with routable IP addresses and DNS means that you can run an in-house Web server. You can run a public Web server for Internet users or a small private Web site to support people in your organization. Running your own Web server enables you to do more things to your site than you can with a Web hosting service, but it also demands more management and development skills.

Two classes of Web server software are available: those that run on Windows NT Server and those that run on Windows 95 and 98. Microsoft's Internet Information Server (IIS) runs on Windows NT Server, and Microsoft's Personal Web Server (PWS) runs on Windows 95, 98, and NT Workstation. Both IIS and PWS are free from Microsoft. If you want to run a serious Web server on Windows 95, 98, or NT Workstation, you should consider a more robust Web server program from Netscape or O'Reilly Software.

Chapter 8 goes into more detail on running your own Web server.

Add Show and Tell to Internet Communications

Net-based video conferencing adds exciting show-and-tell capabilities to Internet communications to enhance brainstorming and collaboration. Video conferencing for face-to-face meetings over DSL can save money and time by reducing travel and can help teleworkers stay in the loop at the office. Video conferencing can be used also for remote expert support, customer service, recruitment, distance learning and training, and telemedicine. The downside of video conferencing is that it demands a lot of bandwidth to get an acceptable quality. A high-quality face-to-face meeting through IP video conferencing begins at a 384-Kbps symmetrical connection.

Video conferencing is a relatively mature TCP/IP application in terms of standards, which go a long way towards ensuring interoperability between products from different vendors — although some incompatibilities do exist. Desktop video conferencing (DVC) systems are relatively inexpensive. And the leading software packages used with DVC hardware are free for the downloading or can be purchased for a nominal cost. These software packages integrate other collaboration tasks into video face-to-face meeting features, such as the capability to share documents and to share ideas using an electronic whiteboard.

Video conferencing and always-on DSL are a great match. If you have an Internet access account with public routable IP addresses, anyone on your LAN (local area network) can use video conferencing for making or receiving video calls. The IP address and a subdomain name act as an always-on telephone number for receiving calls. A video conference participant calls the IP addresses using his or her video conferencing software to make a connection to the receiver.

Telework in the VPN Fast Lane

Teleworkers include telecommuters or anyone who uses information and telecommunications technologies to replace travel. DSL and virtual private networking (VPN) make telecommuters feel like they're sitting at desks in the corporate office, even though they're working remotely. In a virtual private network, you connect to another computer over the Internet and use encryption to protect the data transmissions going on between the two computers. VPN enables teleworkers to link up to a company network through the Internet and work securely, as if they were sitting at a LAN client at the office.

VPN offers some compelling benefits for companies. Remote VPN users bypass the telephone system for long-distance dial-up calls and instead connect as a local call through their ISP. For companies, VPN replaces dial-up remote access servers, modem banks, and telephone lines. Instead, remote access users connect to the network through a single high-speed connection.

Use Push and Pull Applications

The Internet is rapidly becoming a broadcasting medium for information and software. Using push and pull technologies, software vendors can send updates to your software automatically. Push and pull technologies can deliver customized news and information directly to any desktop on your LAN. Push and pull applications were made for DSL service because your Internet connection is always on. This means news, stock quotes, and other information can constantly stream down to your office.

Level the Data Communications Playing Field

Small businesses are already heavy users of the Internet, but because of the high cost of traditional dedicated connections (such as T-1 lines), they could not tap into the full power of the Internet. DSL fills the chasm between slower dial-up modem and ISDN services and the fast but expensive T1 and frame relay services. DSL service is squarely targeted at this bandwidth sweet spot. A DSL connection enables smaller businesses to do the same types of things with their Internet connection that large companies do, such as running Web and e-mail servers and supporting VPN (virtual private networking) for telecommuters.

Run Your Own E-mail Server

E-mail service is an essential component of any Internet connection. The benefits of running your own e-mail server include saving on the monthly charges for e-mail box hosting and the capability of offering more sophisticated e-mail services to your LAN users.

Installing and running your own e-mail mail server on your network allows you to manage your Internet e-mail along with your LAN e-mail. If you elect to run your own mail server, mail for your domain is sent directly from the Internet to the mail server on your network. Connections are made directly from the mail sender on the Internet to your mail server; servers at the ISP are never touched. Typical configurations include providing a mailbox for each user locally on the mail server or implementing mail forwarding and redirection of a user's mail to a separate internal or external mail server where the user's mailbox actually resides.

Share Your Internet Connection

The high-speed, always-on nature of DSL service makes it ideal for sharing it across multiple computers using an Ethernet LAN. Today, setting up the basic LAN plumbing of network interface cards (NICs), cabling, and hubs or switches is easy and inexpensive. Using a single DSL line, you can eliminate using a separate modem and POTS line for each computer on a LAN. As a bonus, for most DSL flavors you can upgrade your bandwidth capacity as traffic increases — without buying a new service or equipment.

A variety of affordable DSL bridges and routers enable you to easily share your DSL service. In most cases, you won't even have to set them up; they're installed as part of your DSL service. Even if you use a DSL service that supports only a single computer, you can use affordable proxy server software or a device called an Ethernet-to-Ethernet router to share the DSL connection.

Tap into Flat Rate Service

DSL shares the same always-on, flat rate qualities of dedicated leased lines but at dramatically lower prices. Because DSL data service is always on, it typically doesn't have usage-sensitive pricing, which means the DSL connection can be used any time for as long as you need without incurring usage charges for the use of the DSL line. The Internet access component of your DSL service, however, might have usage-based pricing after a certain threshold of data (typically 10 gigabits) passes through the connection.

Chapter 15

Ten DSL Service Shopping Tips

*T*he DSL service terrain is a maze. To tap into the promise of high-speed, always-on Internet connectivity, you need to be an informed DSL consumer. What you don't know can cost you both in money and time. This chapter gives you some pointers to help you get the best DSL service for your needs and budget.

Use the Caveat Emptor Defense

The consumer adage *caveat emptor* (buyer beware) is alive and well when it comes to shopping for DSL service. These are turbulent times for DSL deployment, service options, CPE options, and prices. Because DSL service is packaged by ISPs, each with its own approach, no single benchmark exists for the service. Hype abounds, and just below the surface lies a host of fine-print traps waiting for the unsuspecting.

You need to do your homework before shopping for DSL service so that you can ask the right questions. Remember, the stakes are high. The same DSL service can vary by hundreds of dollars per month from one ISP to another!

Shop around to Compare DSL Offerings

Depending on where you're located, you may have a number of ISPs offering DSL service. Behind these ISPs may be multiple DSL circuit service options including both CLECs and ILECs. There may even be an ISP in your area that acts as a CLEC delivering its own DSL service. If you're lucky enough to be in an area serviced by multiple DSL circuit providers, a variety of options will be available. Take advantage of your good fortune and do your comparative shopping.

Comparative shopping for DSL service is a complex endeavor that requires looking beyond just the price. You need to compare apples to apples, not apples to oranges. If you're in an area where both CLECs and ILECs offer DSL service, you have access to different service options. An ILEC's ISP might offer a low-cost ADSL or UADSL service, whereas a CLEC's SDSL offering is more expensive. Note, however, that there could be differences in what you're getting. The ILEC's ISP might offer a dynamic IP addressing service which, however, doesn't support multiple users and doesn't allow you to use your own domain name with the service. You can get around this limitation by using a proxy server or an Ethernet-to-Ethernet router to share the connection and an e-mail and Web hosting service to add your domain name to e-mail service and a Web server — but it will cost more. The CLEC's DSL offering typically includes a multiuser bridge and router as well as support for DNS service.

Three types of ISPs offer DSL-based Internet access service:

✔ **Independent ISPs.** These ISPs buy their DSL circuits from CLECs or ILECs and provide the bulk of DSL service offerings. In large metropolitan markets, independent ISPs typically buy DSL service from multiple DSL circuit providers. For example, an ISP serving Northern California might offer DSL service from Pacific Bell, NorthPoint, and Covad Communications.

✔ **ILEC ISPs.** These ILEC-owned ISPs provide an Internet access package added to the ILEC's DSL offerings. ILEC ISPs have names like PacBell.net, BellAtlantic.net, and US WEST.net. They compete with independent ISPs, but most focus on offering only dynamic IP address accounts targeted at the consumer market.

✔ **ISPs acting as CLECs**. These ISPs become CLECs by filing tariffs with the state regulatory agency, installing DSLAMs in COs, and using the ILECs' local loops in a similar way that larger CLECs do. One of the pioneers of this approach is Harvard Net, which is located in the Boston area.

Know What You Need Before Ordering

Before you make the call to the ISP to start talking turkey about their DSL service, you should have a good idea of what elements you want as part of your DSL service. You need to ask yourself a bunch of questions about what you want, including:

✔ What TCP/IP applications do I want to use?

✔ What is the IP address configuration I want to use?

✔ What will my current and future bandwidth needs be?

✔ Do I want to connect a single computer or multiple computers?

✔ Do I want bridged or routed service?

✔ What are my security options?

Chapter 16 goes into more specifics on IP networking and TCP/IP security considerations.

Don't Be Afraid to Ask Questions

The people you're talking to at the other end of the telephone line at the ISP may or may not have all the answers or present you with all the details. The ISP salespeople are in the business of selling you the DSL service and value-added services. You might need patience and persistence to get the specific answers you want.

Know Your CPE Options and Prices

DSL CPE plays an integral role in the DSL service package because it defines what you can and can't do with your DSL connection. For example, a DSL provider might offer a high-speed DSL Internet access package that looks great — until you discover that the DSL CPE they're using restricts the service to a single computer. Although you can get around this limitation, it involves buying and setting up more hardware or software.

Remember, the differences in CPE technologies — such as PCI adapter and USB modems, single-user bridges, LAN bridges, and routers — have a big effect on your DSL service.

Ask about Promotional Deals

Because DSL service is new, many DSL providers and ISPs are offering special deals, which can save you some cash. For example, an ISP may offer free installation or a discount on DSL CPE. Many times these promotions are offered by the DSL circuit provider and passed on to the customer through the ISP. Ask for any deals the ISP may be offering and check ISP sites for any special DSL service specials.

You don't have to go any further than this book to find special deals. Check out the coupons in the back — they can save you hundreds of dollars in getting your DSL service.

Ask for Price Protection Guarantees

Many DSL service packages are contracted for anywhere from one year to three years, with penalties for early termination. The ISP might offer a better price break for a longer-term contract, but there are dangers with long-term commitments for DSL service. The biggest problem with a long-term commitment for DSL service at a fixed price is that DSL service pricing trends are generally downward. If prices drop for DSL service, do you want to be stuck paying the older, higher rate? Ask the ISP for a price protection deal that allows your DSL price to go down if they lower their prices.

Get a Written Quote

Shopping for DSL means picking up the telephone and calling the ISP to work out your DSL service package. After you've discussed what you want and what it will cost, get a written quote to confirm your verbal discussion. An ISP's quote should provide the following information:

✔ The DSL service and speed you specified

✔ The CPE used for the service and its cost

✔ The breakdown of one-time setup charges

✔ Total monthly recurring charges

✔ Additional charges for custom services, IP addresses domain services, and any additional e-mail boxes

✔ A copy of the terms and conditions of your DSL service

Read the Fine Print

Most ISPs have a terms and conditions document that spells out what you can and can't do with your DSL service. These contracts are important and you should read them carefully. The terms and conditions in this DSL service contract can make or break a DSL service deal.

Restrictions on your DSL service are spelled out in this contract. Here are some important things to look for in a DSL service contract:

✔ **What is the time commitment for your DSL service?** Many ISPs require that you make at least a one-year commitment to the DSL service, with some ISPs requiring a two- or three-year commitment. Depending on the provider, the cost per month may drop if you commit for a longer period. Before committing to a longer-term contract, consider that prices will probably come down as competition heats up. Related to the time commitment is checking for early termination charges.

✔ **Are there usage restrictions?** Many ISPs include restrictions on the amount of data going through your DSL pipe per month per line. For example, there might be a 10-gigabyte limit for a lower-bandwidth DSL connection. Any data moving across your DSL line in excess of the limit costs extra, usually based on a price per megabyte. You may find an even harsher restriction that forbids you from running any servers or using a router that supports NAT and DHCP to share the connection.

✔ **What are the terms of payment?** Increasingly, many ISPs are billing to credit cards to cut down on their accounts receivable overhead. Others will bill you on a monthly or annual basis, with the annual billing method offering a discount. Many ISPs also require a deposit to start the service.

Ask about Backbones and QoS

The bandwidth you're buying to access the Internet through DSL is really the speed supported by the DSL line from your premises through the DSLAM (Digital Subscriber Line Access Multiplexer) at the CO (central office) to the DSL provider's (CLEC or ILEC) backbone. A *backbone* is a segment of the network used to connect smaller segments of networks together. If the connections between the ISP and the DSL provider's backbone and the ISP to the Internet are too small to support the volume of traffic that a number of DSL users causes, a bottleneck occurs.

The ISP business is about leveraging bandwidth to get the most bang for the buck. The urge to oversubscribe is a compelling economic issue with ISPs, and they all play the bandwidth game. Ask the ISP about their backbone network capacity. Note, however, that what they tell will be difficult to verify until you actually get the service. An ISP that charges more for DSL service might do so because they maintain enough backbone to support their DSL subscribers.

Most DSL service is sold as a best-effort service, which means that the DSL circuit provider and ISP do their best to deliver the service but you have no guarantees. As DSL matures, you'll see more Quality of Service (QoS) DSL offerings which guarantee a certain level of performance. A QoS offering is a premium service that costs more.

Chapter 16

Ten Questions to Ask about DSL Service

● ●

In This Chapter

▶ Finding out about the availability of DSL

▶ Discovering your bandwidth requirements

▶ Determining the type of CPE you want to use

▶ Uncovering your IP configuration

▶ Coming up with a strategy for TCP/IP applications

▶ Computing the cost of your DSL service

▶ Finding a nurturing ISP

▶ Unearthing terms and restrictions

▶ Connecting more than one computer

▶ Figuring out what other services you want from your ISP

● ●

*T*he process of establishing a DSL connection involves evaluating interrelated components that together make up your complete DSL service package. This chapter contains the key questions you'll want answered as part of checking out your DSL Internet service options.

Is DSL Service Available in Your Area?

You may be ready for DSL, but DSL may not be ready for you. DSL deployment is on a CO-by-CO basis, which results in an uneven patchwork of DSL service availability.

If you live in a large metropolitan area with high concentrations of potential DSL customers, chances are that an ILEC, a CLEC, or an ISP acting as a CLEC is offering DSL service in your area. You may even be in an area where you can choose a DSL service among all three. If you live in a smaller city, a town, or a rural area, your chances of having access to DSL service get smaller.

If you live in a newer housing development or work in an office park, a Digital Loop Carrier (DLC) might service the area. DLCs are typically used in office parks and housing developments to minimize the need to run local loops over several miles to the CO servicing the area. DLCs cover about 25 percent of the local loops in the United States. For some ILECs, DLCs cover closer to 50 percent. Unfortunately, DLCs were designed for POTS, not for DSL service. Solutions to this problem, however, are being deployed.

Finding out about DSL service availability is a fragmented process that over time will improve as large-scale DSL availability databases are put online. In the meantime, the best starting points for checking DSL service availability are the ILEC and CLEC Web sites. Chapter 9 provides a detailed listing of the major ILECs and CLECs offering DSL service and their ISP partners.

If you know ISPs in your area, check out their Web sites. You can also check out Internet.com's "The List," which provides the most comprehensive database of ISPs at `thelist.internet.com`. Another good site for ISP listings is the `www.xdsl.com` site operated by TeleChoice, a telecommunications consulting company.

What's Your Need for Speed?

In the real world, the biggest constraint on bandwidth is cost. The more bandwidth you want, the more it will cost. Bandwidth capacity planning is one of the most important and most difficult tasks you'll undertake in setting up a DSL connection. A number of factors come into play when trying to evaluate your bandwidth needs. Chapter 7 goes into more detail on estimating bandwidth requirements.

Here are some basic bandwidth questions you need to answer:

- **Will you be running any servers on your DSL connection?** A big factor in determining your bandwidth requirements is whether you plan to run any servers on your DSL connection. If you plan to run a Web server, for example, you need to take into consideration the incoming traffic as well as your outgoing traffic. Chapter 8 provides formulas to calculate bandwidth needs for running a Web server.

- **Do you plan to use IP voice and video conferencing?** If used even moderately, these cool but bandwidth-hungry applications can eat up your DSL connection. Supporting multiple simultaneous users compounds the demands.

✔ **Do you plan to connect one computer or a network to your DSL line?**
The more PCs sharing the DSL connection, the greater the demand for
bandwidth.

✔ **Can you upgrade to a higher speed?** One of the best features of most
(but not all) DSL service offerings is their bandwidth scalability. This
means you can upgrade to a higher speed over time without having to
start over again with new CPE.

Which Type of CPE Do You Want to Use?

What stands between you and the Internet through DSL is the Customer
Premises Equipment (CPE). The type of DSL CPE you use as part of your DSL
service plays an integral role in defining your Internet connection capabilities. Because DSL offerings are typically sold as a complete package that
includes the DSL service, Internet access, and CPE, you need to understand
the differences in DSL CPE offered as part of a package. Related to your CPE
decision are IP configuration considerations, which are explained next in the
"What's Your IP Configuration?" section.

DSL CPE devices break down into single-user or multiuser solutions. Single-
computer DSL CPE consists of the following:

✔ PCI (Peripheral Component Interconnect) adapter cards

✔ USB (Universal Serial Bus) modems

✔ Bridges

These single-user solutions are usually bundled with dynamic IP Internet
access accounts from the ISP. This means that the IP address changes
depending on the lease times established by the ISP. Most dynamic IP
Internet service doesn't support domain name service, which means you
can't use your own domain name as part of the service. Many ILEC, ADSL,
and UADSL (G.Lite) offerings from their in-house ISPs use dynamic IP
addressing.

Dynamic IP Internet access is usually bundled with CPE that restricts access
to only a single computer. Most of these single-user DSL CPE options can be
modified to support multiple users by using proxy server software or
Ethernet-to-Ethernet routers. Proxy server software and Ethernet-to-Ethernet
routers also provide firewall protection for your LAN. Chapter 6 goes into
detail on proxy servers and Internet security issues.

DSL CPE for the multiuser network environment consists of the following:

- ✓ LAN modems (bridges)
- ✓ Routers

DSL bridges are often referred to as DSL LAN modems. Unlike the single-user bridge, a DSL LAN modem connects to a hub, which makes the DSL service available to all the computers connected to the hub or switch. In actuality, many LAN modems available today are hybrid devices that incorporate bridge as well as some router capabilities. Most of these LAN modems, however, don't support as many users as routers, and they don't support NAT and DHCP.

A router is a more sophisticated gateway device than a bridge. A router allows data to be routed to different networks based not on hardware addresses (as in a bridge) but on packet address and protocol information. This decision-making functionality, called filtering, not only enables a router to protect your network from unwanted intrusion but also prevents selected local network traffic from leaving your LAN through the router. This is a powerful feature for managing incoming and outgoing data for your site. Routers typically cost about twice as much as bridges.

See Chapter 5 for more information on all DSL CPE options, including Ethernet-to-Ethernet routers.

What's Your IP Configuration?

The type of IP addressing you use as part of your DSL Internet service plays a pivotal role in determining the kind of interaction you have with the Internet. IP addressing defines what you can and can't do with your Internet connection. A dynamic IP address configuration is targeted to the consumers and small businesses that don't plan to run TPC/IP applications that require a static IP address, such as any kind of Internet server or Net voice or video conferencing applications. However, you can share a dynamic IP Internet account using a router with NAT, proxy server software, or an Ethernet-to-Ethernet router, and you can use a Web and e-mail hosting service with your own domain name.

Using static IP addresses means using IP addresses that are recognized and routable on the Internet. Businesses and power users typically use this type of IP addressing. Static IP address Internet access accounts cost more because you must lease IP addresses from the ISP. They also require more configuration tasks for your computers and CPE. IP addresses linked to specific hosts and domain names enable Internet users to access a host computer running as a Web server (or any TCP/IP application server) using a user-friendly text identifier, such as www.angell.com. This type of IP addressing can be used to make you a provider of Internet services for all Internet users or for a private workgroup.

Bridged, or routed, DSL service requires three IP addresses just for the IP server. One IP address is for the router, one is for the Ethernet connection, and one is for the WAN connection. If you get a block of eight IP addresses, for example, only five are available for hosts on your LAN.

Blocks of IP addresses are available from most ISPs for a monthly cost based on the number of IP addresses. Some ISPs include a block of IP addresses as part of the service. The IP addresses assigned to you by the ISP are available for use while you are the ISP's DSL customer. They remain the property of the ISP and return to the ISP upon termination of the service.

The ISP also provides name resolution service so that TCP/IP applications can use Internet domain names instead of just numeric IP addresses. The ISP will typically register a domain name on your behalf for no cost, but you will be billed for the domain name directly from Network Solutions. If your domain was previously hosted at another ISP or hosting service and you want to move it to a new ISP, the new ISP will typically do it without charge.

Chapter 4 deals with TCP/IP considerations as part of your DSL service package.

What's Your TCP/IP Application Strategy?

The temptation to try out all kinds of cool, sophisticated client and server applications on your DSL connection is compelling. Stop, and remember that each application includes pros and cons that need to be looked at carefully and understood in terms of their effect on your DSL connection.

Here are some guidelines to keep in mind as you explore your TCP/IP application options and formulate your strategy:

 ✔ Pursue a thoughtful strategy that compares the benefits to the commitment it takes to fully utilize an application. There are always trade-offs between an application's benefits and costs.

 ✔ Be careful if you plan to run any type of server (Web server, e-mail server) on your DSL connection. Running a public Web server on your

LAN, for example, can make big demands on your DSL connection as well as on your time and effort to set up and administer the server.

✔ IP video conferencing is a sexy application that holds a lot of promise, but it's bandwidth hungry and can eat up your DSL connection.

✔ Consider using a remote hosting service for your Web server, e-mail, and other Internet services. Hosting a Web site and e-mail boxes with your domain name is affordable, and lots of packages are available.

What Will DSL Service Cost?

The total price you pay for your DSL service depends on several variables. You need to crunch the numbers from one-time installation charges, CPE costs, monthly service charges, and other additional charges to get to the bottom line of what DSL Internet service will cost.

The two main costs of DSL service are the one-time start-up and CPE charges, and the recurring monthly charges for DSL and Internet service. Start-up costs are where you'll take the biggest hit for DSL service. Start-up costs can include an ISP activation charge, a DSL circuit activation fee, an onsite installation charge, a CPE charge, as well as setup fees for IP addresses, e-mail boxes, and other ISP services. The fixed charges for setting up DSL service doesn't vary much from one speed to another within the same DSL flavor. The biggest variable cost is the monthly service charges for the different speeds.

Monthly recurring costs can include the ISP charge, the DSL circuit charge, the CPE lease cost, and other ISP fees for IP address services, e-mail boxes, and so on. Monthly fees are usually consolidated into a single bill. The two main components of your monthly bill for DSL service are the circuit charge from the ILEC or CLEC and the Internet service cost. The circuit charge is typically part of a single bill from the ISP, but can be on a separate bill when an ILEC is the DSL circuit provider. The Internet service part of your monthly charge might have a base flat rate and a usage charge based on megabytes for data traffic exceeding the limit.

Chapter 7 goes into more detail on estimating the costs of setting up and using DSL service. Chapter 10 tells you all about available ISP services.

How Helpful Is the ISP?

The ISP's Web site is an important starting point for your DSL service shopping. A good Web site should help customers develop their package before they call to talk to the ISP salesperson. Unfortunately, many ISPs lack good DSL customer information at their sites.

Here are some guidelines to judge a helpful ISP Web site:

✔ **Does the Web site provide specifics on the ISP's DSL service offerings?** Glossy marketing copy doesn't make for good consumer information. Information packaged to educate the consumer about the service and product details is helpful.

✔ **Does the Web site include the costs of the ISP's DSL offerings?** Installing and using DSL service involves a variety of charges. A Web site should provide a breakdown of the costs for getting the DSL service, including a menu of optional services.

✔ **Does the Web site include the ISP's terms of the service?** Unfortunately, most ISPs don't post their terms and conditions on their Web sites. These documents are the fine print of your DSL service. Terms and conditions are usually provided as part of the formal quote.

✔ **Does the Web site provide good CPE information?** CPE is at the heart of your DSL service capabilities and defines your TCP/IP application options. The Web site should provide coverage of the different CPE options and their prices.

✔ **Does the Web site provide information on IP addresses and domain name services?** A good Web site lists your IP address options and costs, such as the additional costs for IP addresses and DNS registration. The site should also include a menu of custom services, such as ISP support for running your own e-mail server.

After you get the basic information about an ISP's DSL offerings, the next point of contact with the ISP is to call them. The staff at many ISPs know their stuff and do a good job at helping you, but others don't. How well an ISP handles this process might be an indicator of how good their service will be. See Chapter 7 for more specifics on shopping for DSL service from an ISP.

What Are the ISP's Terms and Restrictions?

The terms and conditions in the DSL service contract can often make or break a DSL service deal. Restrictions on your DSL service are spelled out in this contract, so you must read it carefully to fully understand what you can and can't do with your DSL service.

Here are some important things to look for in a DSL service contract:

✓ **What is the time commitment for your DSL service?** Many ISPs require that you make a one- to three-year commitment to the DSL service. With some providers, the cost per month drops if you commit for a longer period. Before committing to a long-term contract, consider that DSL service is new and prices will probably come down as competition increases. Check also for any early termination charges.

✓ **Is usage restricted?** Many ISPs include restrictions on the amount of data going through your DSL pipe per month per line. For example, there might be a 10-gigabyte limit for a lower-bandwidth DSL connection. Any data moving across your DSL line in excess of the limit costs extra, usually based on a price per megabyte. You may find an even harsher restriction that forbids you from running any servers or using a router that supports NAT and DHCP to share the connection.

✓ **What are the payment terms?** Increasingly, ISPs are billing to credit cards to cut down on their accounts receivable overhead. Others bill on a monthly or annual basis, with the annual billing method offering a discount. Many ISPs also require a deposit to start the service.

✓ **Is a Quality of Service guarantee available?** Most ISPs don't offer any Quality of Service (QoS) guarantees for DSL service at this point. DSL is usually sold as a best effort service. However, most CLECs are offering SLAs (Service Level Agreements) for the connection between the CPE and DSLAM. As DSL matures, you'll probably see a premium business class of DSL service, in which the ISP guarantees a level of service for a higher price.

Chapter 7 goes into more details on what to check out when shopping for DSL service.

Do You Plan to Connect Multiple Computers?

The high-speed, always-on nature of DSL service makes it ideal for LAN-to-Internet connections. The inherent benefit of building a LAN, beyond the sharing of local resources, is the capability to share the high-speed DSL connection across multiple computers. Even if you're using a single-user (dynamic IP address) DSL Internet access service, with a LAN you can share the service across multiple computers by adding a proxy server or an Ethernet-to-Ethernet router.

Today, setting up the basic LAN plumbing for your office or home is easy and inexpensive. Building an Ethernet network from the ground up involves adding network interface cards (NICs) to your computers, and then connecting them with cabling to a network hub or a switch device. Two Ethernet specifications are available for PC networks: standard Ethernet (called 10Base-T) and the newer Fast Ethernet (called 100Base-T). The cost differential for 10Base-T versus 100Base-T networking hardware is becoming inconsequential. Going for Fast Ethernet allows you to buy for your current networking needs and invest in a scalable technology for future bandwidth demands.

What Other Services Do You Want from an ISP?

Beyond the basic DSL and IP networking services, you should consider a variety of related TCP/IP-based services as part of your Internet access package. The two most likely services you'll want from an ISP are e-mail and Web hosting services.

E-mail service is an essential component of any Internet connection. When it comes to getting Internet e-mail service, two options are typically available: hosting your e-mail services with your DSL ISP or operating your own mail server on your network. For smaller organizations, hosting your Internet mailboxes with the ISP makes the most sense.

Installing and running your own SMTP mail server on your network allows you to manage your Internet e-mail along with your LAN e-mail. Configuration and administration of the mail server is your responsibility. The ISP typically charges a fee for configuring their e-mail server to send all your e-mail directly to your e-mail server.

A DSL connection using routable IP addresses and domain name service can support running a Web server on your LAN. You should, however, evaluate the pros and cons of running your own Web server versus using a Web hosting service from the ISP (or from another Web hosting service).

Part V
Appendixes

The 5th Wave   By Rich Tennant

"Just how accurately should my Web site reflect my place of business?"

In this part . . .

This part includes two handy resources to help you in your DSL education. Appendix A is an extensive listing of DSL products, services, and other resources for quick reference. Appendix B provides a useful glossary of the DSL lingo and jargon that you're likely to come across in your journey toward DSL enlightenment.

Appendix A

DSL Products, Services, and Resources

In This Appendix

▶ Vendors of DSL CPE

▶ Proxy server and other security products

▶ TCP/IP utilities and applications

▶ Web sites of the leading ILECs and CLECs

▶ Additional resources for information on DSL

This appendix provides you with contact information for checking out DSL products, services, and information resources.

DSL CPE Vendors

A variety of DSL CPE is available, including PCI and USB modems, DSL bridges, and DSL routers. Table A-1 lists the leading vendors of DSL CPE and their Web site URLs so that you can check out the latest products. Many of these vendors manufacture a complete line of DSL CPE products.

Table A-1	Vendors of DSL CPE
Vendor	*Web Site for Product Information*
3Com	www.3com.com/solutions/dsl/
Alcatel	www.alcatel.com
Cayman Systems	www.cayman.com

(continued)

Table A-1 *(continued)*

Vendor	Web Site for Product Information
Cisco	www.cisco.com
Copper Mountain Networks	www.coppermountain.com
Efficient Networks	www.efficient.com
Escalate Networks	www.escalate.net
FlowPoint	www.flowpoint.com
Netopia	www.netopia.com
Paradyne	www.paradyne.com/
Westell	www.westell.com

DSL Proxy Server and Security Products

A number of vendors offer proxy server software and hardware products. The hardware versions are commonly referred to as Ethernet-to-Ethernet routers. These products allow you to share an Internet connection that uses single-user DSL CPE as part of the DSL service package. Proxy server products also provide the bonus of security for your LAN. A related product called an Internet firewall appliance provides proxy server functions but also includes additional security features. See Table A-2 for details.

Table A-2 Vendors of Proxy Server and Security Products

Vendor	Web Site	Product(s)
Cayman Systems	www.cayman.com	Cayman 2E 500, Cayman 2E 500-H Ethernet-to-Ethernet routers
Microsoft	www.microsoft.com/proxy/	Microsoft Proxy Server; runs only on Windows NT
MultiTech Systems	www.multitech.com	ProxyServer; hardware proxy server (Ethernet-to-Ethernet router)
Netopia router	www.netopia.com	R7100 Ethernet-to-Ethernet
Sonic Systems	www.sonicsys.com	SonicWALL, SonicWALL Plus, SonicWALL DMZ

Vendor	Web Site	Product(s)
WinGate	www.wingate.com	WinGate; proxy server software for Windows 95/98 and NT
WinProxy	www.winproxy.net	WinProxy; proxy server software for Windows 95/98 and NT

TCP/IP Utilities and Applications

The combination of high speed and an always-on DSL-to-Internet connection opens up an intoxicating number of TCP/IP application possibilities. This section provides a sampling.

TCP/IP utilities

Table A-3 lists some popular tools for monitoring the performance of your DSL and Internet connection, as well as other IP networking tools.

Table A-3	Vendors of TCP/IP Utilities	
Vendor	Web Site	Product(s)
Net.Medic	www.vitalsigns.com	Windows tool for monitoring the performance of your Internet connection
NetSwitcher	www.netswitcher.com	A handy Windows 95/98/NT utility for switching between multiple Ethernet TCP/IP connections
Visual Route	www.visualroute.com	A Windows-based IP tracer program

Video conferencing

Video conferencing adds exciting show-and-tell capabilities to Internet communications. Table A-4 shows you the leading video conferencing system vendors.

Table A-4	Vendors of Video Conferencing Products	
Vendor	**Web Site**	**Product(s)**
3Com	www.3com.com	3Com BigPicture; PCI video conferencing system
Diamond Multimedia	www.diamondmm.com	Supra Video Kit; PCI video conferencing system
Kodak	www.kodak.com	Kodak DVC323; USB video conferencing system
Kolban	www.kolban.com/webcam32/	WebCam32; shareware
Microsoft	www.microsoft.com/msdownloads	Microsoft NetMeeting; free video conferencing software that works with most video conferencing systems
White Pine	www.wpine.com	CU-SeeMe video conferencing software; MeetingPoint Web video conferencing server software.
Winnov	www.winnov.com	Videum Conference Pro; PCI and USB video conferencing systems
Xirlink	www.xirlink.com	C-it; USB video conferencing system
Zoom Telephonics	www.zoomtel.com	ZoomCam; USB video conferencing system

Web servers

A high-speed, always-on DSL connection combined with routable IP addresses and DNS means you can run your own in-house Web server. In Table A-5, you can check out the leading Web server software programs.

Table A-5	Vendors of Web Server Programs	
Vendor	**Web Site**	**Product(s)**
Apache	www.apache.org	Apache HTTP server for Microsoft Windows
Microsoft	www.microsoft.com	Internet Information Server (Windows NT); Personal Web Server (Windows 95/98)

Vendor	Web Site	Product(s)
Netscape	www.netscape.com	FastTrack Server and Enterprise Server (both run on Windows NT)
O'Reilly Software	Website.oreilly.com	Web Site Professional

E-mail servers

E-mail service is an essential component of any Internet connection. The benefits of running your own e-mail server include saving on monthly charges for e-mail hosting and being able to offer more sophisticated e-mail servers for your LAN users. Table A-6 lists the leading e-mail server programs.

Table A-6   Vendors of E-mail Server Programs

Vendor	Web Site	Product(s)
deerfield.com	www.mdaemon.com	MDaemon Mail Server for Windows (Windows 95/98/NT)
Eudora	www.eudora.com/worldmail	WorldMail Server (Windows NT Server/Workstation)
Microsoft	www.microsoft.com	Microsoft Exchange Server (Windows NT Server)
Netscape	www.netscape.com	Netscape Messaging Server (Windows NT Server)

Net phones, net faxing, and IP paging

Net phones, net faxes, and IP pagers bring popular communications tools to the Internet. Net phone applications allow you to make and receive voice calls over the Internet. Net phones and video conferencing are closely related, and in some cases are combined into a single application, such as Microsoft NetMeeting.

Net-based faxing uses software and a fax provider service to route faxes to and from the Internet to fax machines connected to the PSTN or as e-mail messages to other Net users. IP paging uses a simple IP pager program running on your system and other users' systems on the Internet, and acts like the pagers used in telecommunications. Check out Table A-7 for a list of vendors.

Table A-7	Vendors of Net Phones, Net Faxing, and IP Paging Programs
Product	**Vendor Web Site**
Net Phone	
Internet Phone	www.vocaltec.com
Net2Phone	www.net2phone.com
NetMeeting	www.microsoft.com
PhoneFree	www.phonefree.com
VDOPhone	www.vdo.net
Net Faxing	
Faxaway	www.faxaway.com
FaxSav	www.faxsav.com
Faxscape	www.faxscape.com
FaxStorm	www.netcentric.com
JFAX	www.jfax.com
IP Paging	
ICQ	www.icq.com

Remote control programs

Remote control software enables a remote client computer to take control of another computer (called a host) through an Internet connection. Table A-8 lists the leading remote control software products.

Table A-8	Vendors of Remote Control Programs
Product	**Vendor Web Site**
Carbon Copy	www.compaq.com/products/networking/software/carbon copy
pcAnywhere	www.symantec.com
Timbuktu Pro	www.netopia.com

Virtual private networking

Virtual private networking (VPN) lets telecommuters connect to their corporate networks via the Internet and work securely remotely. Some of the leading VPN vendors are listed in Table A-9. Also check DSL router vendors. Many now offer VPN support built into the router.

Table A-9	Vendors of Virtual Private Networking Programs	
Vendor	*Web Site*	*Product(s)*
Indus River Networks	www.indusriver.com	RiverWorks Tunnel Server, RiverWorks Management Server, RiverMaster, RiverPilot, and RiverWay
Microsoft	www.microsoft.com	PPTP (Point-to-Point Tunneling Protocol), a software solution included with Windows NT (VPN server) and Windows 95/98 and NT Workstation (VPN client)
RedCreek	www.redcreek.com	Personal Ravlin and Ravlin family of VPN products
VPNware Systems	www.vpnet.com	VPNware, VPNremote, VPNmanager

ILEC and CLEC Web Sites

Tabel A-10 represents the Web sites for the leading ILEC and CLEC DSL providers. These Web sites are a good place to start your DSL service availability search.

Table A-10	Web Sites of the Leading ILECs and CLECs
ILEC/CLEC	*Web Site*
Ameritech (ILEC)	www.ameritech.com/products/data/adsl/
Bell Atlantic (ILEC)	www.bell-atl.com/infospeed/
Bell South (ILEC)	www.bellsouth.net/external/adsl/
Covad Communications (CLEC)	www.covad.com
GTE (ILEC)	www.gte.com
NorthPoint Communications (CLEC)	www.northpointcom.com
Pacific Bell (ILEC)	www.pacbell.com/products/business/fastrak/adsl/
Rhythms Net Connections (CLEC)	www.rhythms.com
Southwestern Bell (ILEC)	www.swbell.com/dsl/
US WEST (ILEC)	www.interprise.com

Other DSL Resources

Table A-11 provides a list of some organizations and resources involved in DSL issues.

Table A-11	Resources for DSL Information	
Resource	*Web Site*	*What It Is*
ADSL Forum	www.adsl.com	One of the leading industry organizations working to develop consensus on DSL standards. The ADSL Forum represents most telecommunications and computer companies involved with DSL services.
C-LEC	www.clec.com	A resource of CLEC news and other information.
DSLCon	www.dslcon.com	An organization hosting DSL conference and trade shows in Dallas, Reston, San Jose, Boston, Rio de Janeiro, and Brussels. Excellent trade shows with good conferences.
DSL Prime	www.dslprime.com	Good resource for what's happening in the DSL industry.
FCC	www.fcc.gov	Federal Communications Commission. The United States agency in charge of telecommunications oversight, policy, and regulation at the federal level.
HISAC	www.hisac.org	High-Speed Access Coalition. A California telecommunications consumer advocacy group.
xDSL.com	www.xDSL.com	A DSL resource site operated by TeleChoice, a telecommunications consulting company.

Appendix B

Glossary

• •

100Base-T The newer Ethernet networking standard that supports a data transmission rate of 100 Mbps and is backward compatible to 10Base-T networks. 100Base-T is based on the IEEE 802.3u standard and is commonly referred to as Fast Ethernet.

10Base-T The Ethernet networking standard that supports a data transmission rate of 10 Mbps. 10Base-T is based on the IEEE 802.3 specification.

2B1Q Two Binary, One Quaternary. A line coding technique used in traditional telecommunications offerings including ISDN. The 2B1Q line coding is used for some DSL flavors, including SDSL, HDSL, and IDSL.

ADSL Asymmetric Digital Subscriber Line. A DSL flavor that supports high-speed data communications speeds up to 8 Mbps downstream and up to 640 Kbps upstream. ADSL can deliver simultaneous high-speed data and POTS voice service over the same telephone line. ADSL is the most widely deployed flavor of DSL by the ILECs.

ADSL Forum An organization made up of computer and telecommunication companies that defines DSL standards for submission to standards bodies. This group is responsible for accelerating ADSL technologies, products, and services as well as promoting the technology.

ADSL Lite A DSL flavor based on the new G.Lite standard that supports 1.5 Mbps downstream and 348 Kbps upstream. ADSL Lite is commonly called Universal ADSL (UADSL) because it's expected to become the most widely deployed DSL flavor.

ANSI American National Standards Institute. The organization that defines standards, including network standards, for the United States.

ATM Asynchronous Transfer Mode. A standard for high-speed digital backbone networks. ATM networks are widely used by telecommunications and large companies for backbone networks that consolidate data traffic from multiple feeders (such as DSL lines) and different types of media (voice, video, and data).

attenuation The reduction of a signal's power as it passes through most media. Usually proportional to distance, attenuation is sometimes the factor that limits the distance a signal may be transmitted through a medium before it can no longer be received. In the high frequencies of DSL service, attenuation plays a big role in the distance limitations inherent in DSL service.

ATU-C ADSL Transceiver Unit-Central office. Equipment placed at the central office to support DSL services. This term has been eclipsed by the term DSLAM.

ATU-R ADSL Transceiver Unit-Remote. Equipment placed at the customer premises to support DSL service. This term has been eclipsed by the more generic term CPE.

AWG American Wire Gauge. A thickness measurement for copper wiring. The heavier the gauge, the lower the AWG number and the better the quality of the line in terms of supporting a longer distance for a DSL signal. Many local loops use 24 AWG or 26 AWG copper wires.

backbone A major transmission path used for high-volume network-to-network connections. In DSL-to-Internet connections, a backbone network consolidates data traffic from the individual DSL lines into a backbone network for delivery to the ISPs.

bandwidth The amount of data that can flow through a given communications channel. It's the difference in the lowest and highest frequency of a data communications channel. The greater the bandwidth, the more data that can travel at one time.

best effort Internet access service that doesn't have a Quality of Service (QoS) guarantee. Most DSL service is a best effort class of service, but QoS is emerging as a premium DLS service by some ISPs.

binary A number system based on 2. The place columns of the number are based on powers of 2: 1, 2, 4, 8, 16, 32, 64, 128, 256, and so on. The binary to decimal conversions make up the IP addresses used on any TCP/IP network, such as 199.232.255.113.

Bps Bits per second. The unit of measurement for data transmission speed over a data communications link.

bridge A device that connects two networks as a seamless single network using the same networking protocol. DSL LAN modems are typically bridges. Bridges operate at the hardware layer and don't include IP routing capabilities.

bridged tap An extension to a local loop generally used to attach a remote user to a central office switch without having to run a new pair of wires all the way back. Bridged taps are fine for POTS, but severely limit the speed of digital information flow on the link.

cable binder A bundle of local loop wires that runs along telephone poles or underground from the CO.

CAP Carrierless Amplitude Phase. A modulation transceiver technology used in ADSL systems.

CAT5 Category 5 unshielded twisted-pair wiring commonly used for 10Base-T and 100Base-T Ethernet networks and rated by the EIA/TIA.

channel A path for digital transmission signals. Within digital services such as DSL, multiple channels can share the same pair of wires. Channels are created using multiplexers.

CIDR Classless Internet Domain Routing. In response to the limitations of A, B, and C classes of IP addresses, InterNIC implemented CIDR (which is pronounced "cider"). CIDR allows IP addresses to be broken down into smaller subnets than the class C network, with 256 IP addresses. CIDR networks are described as slash x networks, where x represents the number of bits in the IP address range.

circuit A path through a network from source to destination and back. In a circuit-switched network, this path uses a fixed route and a fixed amount of bandwidth for the duration of the connection between the two end points.

CLEC Competitive Local Exchange Carrier. A competitor to ILECs offering telecommunication service. In the case of DSL service, the CLECs offer data communications service.

client A program or a device that requests services from a server.

client/server A style of computer networking that allows work to be distributed across powerful computers acting as servers and client computers. TCP/IP uses a client/server architecture.

CO Central office. A telephone company facility within which all local telephone lines terminate and which contains the equipment required to switch customer telecommunications traffic. For DSL service, DSLAM equipment is typically set up at the CO to support DSL service for lines terminating at the CO.

CPE Customer Premises Equipment. A telecommunications term that refers to any equipment located at the customer's premises. DSL modems, bridges, and routers are CPE.

crosstalk The interference induced on a signal on one line that is caused by the transfer of energy from a colocated line. Crosstalk is a factor in the delivery of different flavors of DSL service in the same cabling bundle.

CSMA/CD Carrier Sense Multiple Access/Collision Detection. A network transmission scheme by which multiple network devices can transmit across the cable simultaneously. CSMA/CD is used as the basis of Ethernet networks.

Data CLEC A Competitive Local Exchange Carrier focusing on IP data communications links and not providing traditional voice telecommunications.

default gateway The address that the IP uses if the destination address is not on the local subnet. It's usually the router's IP address.

demarcation point The point at the customer premises where the line from the telephone company meets the premises wiring. From the demarcation point, the end user is responsible for the wiring.

DHCP Dynamic Host Configuration Protocol. A protocol that allows IP addressing information to be dynamically assigned by a server to clients on an as-needed basis. IP addresses for a network are stored in a pool of available IP addresses, which are allocated when a computer on the network boots up. The DHCP server functionality is built into most DSL routers.

DLC Digital Loop Carrier. A telecommunications structure that connects end users located more than 18,000 feet (3.5 miles) away from the central office. DLCs consist of a box containing line cards that concentrate individual lines within a given area and then send the traffic over a high-speed digital connection. DLCs are commonly deployed in office parks and residential subdivisions.

DMT Discrete Multi-Tone. An ADSL modulation technique standardized by the ANSI T1E1.4 standard and used in ADSL systems.

DNS Domain Name System. The name resolution service for IP addresses that provides the friendlier text-based addresses for Internet resources. DNS uses a distributed database containing FQDNs and addresses.

domain name A group of computers using the same DNS name servers and managed within the same administrative unit.

domain name server A program that converts FQDNs into their numeric IP addresses, and vice versa.

Domain Name Service The configuration of user-friendly text domain names to IP addresses by an ISP using the Domain Name System.

downstream The direction of data flow on a data communications link, which occurs from the network down to the user. In the case of Internet access, it's the capacity or speed of data flowing from the Internet to the end user's PC or LAN.

DSL Digital Subscriber Line. The generic term that refers to the underlying technology inherent in all flavors of DSL, such as ADSL, SDSL, and HDSL.

DSLAM Digital Subscriber Line Access Multiplexer. The device typically housed at the CO that terminates all the DSL lines serviced by the CO. The DSLAM consolidates or concentrates all the data traffic coming in from individual DSL lines and passes them on to a backbone network for distribution to Internet service provider networks or corporate networks.

dynamic IP addressing An IP address is assigned to the client for the current session or some other specified amount of time. This form of IP addressing is used by DSL services targeted at the consumer market and typically doesn't support any Domain Name System service.

EIA/TIA Electronic Industries Association/Telecommunications Industry Association. An organization that provides standards for the data communications industry, including the cabling used for networking and telecommunications.

Ethernet A LAN technology that uses CSMA/CD delivery and can run over different media (cabling). Most of today's Ethernet LANs use twisted-pair 10Base-T wiring that can support both standard Ethernet at 10 Mbps and the newer Fast Ethernet at 100 Mbps.

Ethernet address The unique hardware address that identifies any Ethernet device, including network interface cards (NICs), network printers, DSL bridges, and routers.

Fast Ethernet The Ethernet-based networking protocol that supports up to 100 Mbps capacity. Commonly referred to as 100Base-T networking.

FCC Federal Communications Commission. The United States government agency for regulating the telecommunications industry.

fiber optics A technology in which light is used to transport large amounts of data using thin filaments of glass. Fiber-optic transmission is used for backbone networks.

firewall A security device (hardware or software) that controls access from the Internet to a local network by using identification information associated with TCP/IP packets to make a decision about whether to allow or deny access. This decision is based on a set of defined rules that describe which packets or sessions are allowed.

FQDN Fully Qualified Domain Name. The full name of a host, including all sub-domain and domain names, separated by dots. For example, `david.support.angell.com` is an FQDN.

fractional T-1 Any data transmission rate between 56 Kbps and 1.54 Mbps (which is the full T-1 rate). Fractional T-1 is simply a digital, dedicated line that's not as fast and not as expensive as a T-1 line.

frame relay A dedicated, public data networking service offered by telecommunication companies for LAN-to-LAN connections. Frame relay uses variable-length frames for packet-switching networks that efficiently handle bursty communications by quickly adjusting bandwidth to meet demands.

gateway A functional device allows equipment with different protocols to communicate with each other. The gateway device can be embodied in a router or a computer.

G.Lite The new ITU standard that supports 1.5 Mbps downstream and 384 Kbps upstream.

hardware address The physical address for the NIC, which is used by low-level hardware layers of the network, including DSL bridges. Also called the MAC address.

HDSL High-bit-rate Digital Subscriber Line. The DSL service widely used for T-1 lines. HDSL uses four wires (two pairs) instead of the standard two wires used for other DSL flavors. HDSL supports symmetrical service at 1.54 Mbps but does not support POTS.

HDSL-2 High-bit-rate Digital Subscriber Line-2. The ITU has approved a new generation of HDSL that offers several enhancements over its predecessor. One of the most important is that HDSL-2 requires only a single twisted-pair local loop instead of the two pairs required for HDSL.

host A computer or any device connected to a TCP/IP network.

hub A passive network device that repeats all data traffic to all ports. A hub is at the center of a LAN, and all networked devices, including computers, printers, and DSL bridges or routers, are connected to the hub through cables.

ICMP Internet Control Message Protocol. The TCP/IP protocol used to report network errors and to determine whether a computer is available on the network. The ping utility uses ICMP.

IDSL ISDN Digital Subscriber Line. The always-on cousin of dial-up ISDN. IDSL delivers a symmetric 144 Kbps of bandwidth, which is 16 Kbps more than the dial-up version of ISDN. This 16-Kbps difference comes from the elimination of the two 8-Kbps channels used in ISDN for communicating with the PSTN switch. Unlike ISDN, IDSL doesn't support POTS.

IEEE Institute of Electrical and Electronics Engineers. A worldwide engineering and standards-making body for the electronics industry. The standards committee for LAN technologies. Advises on standards for ANSI.

IEEE 802.3 The local area network protocol known as Ethernet. The 802.3 protocol forms the basis for 10-Mbps or 100-Mbps throughput that uses CSMA/CD. This allows LAN users to share the network cable, but only one station can use the cable at a time.

IETF Internet Engineering Task Force. The organization that provides the coordination of standards and specification development for TCP/IP networking. Part of the IAB (Internet Architecture Board) responsible for research into Internet issues. RFCs (Request For Comments) document the IETF specifications.

ILEC Incumbent Local Exchange Carrier. A new term that emerged from the Telecommunications Act of 1996 that describes the traditional local telephone companies, which control local telephone service (voice or data). Companies competing with these ILECs are called CLECs.

IMAP4 Internet Message Access Protocol, Version 4. IMAP4 provides sophisticated client/server capabilities beyond the features of POP3. POP3 and IMAP4 don't interoperate, but many e-mail servers and clients can support both protocols.

Internet address The unique 32-bit numeric address, such as 199.232.255.113, used by a host on a TCP/IP network. The IP address consists of two parts: a network number and a host number.

interoperable Two pieces of equipment are interoperable when they work together. Standards are designed to enable interoperability among different devices from different vendors.

intranet A local network that uses TCP/IP and Web technologies as its networking protocol and information resource interface. Internal company information made available using the Web browser or other TCP/IP applications. Many intranets are protected from exterior access by various security devices, such as routers, proxy servers, or firewalls.

IP Internet Protocol. The connectionless network layer protocol that forms the networking functions of the TCP/IP suite. IP networking forms the basis of networking over the Internet and allows information to be transmitted across dissimilar networks.

IP address Internet Protocol address. A 32-bit dotted decimal notation used to represent IP addresses. Each part of the address is a decimal number separated from other parts by a dot (.), such as 199.232.255.113.

IPSec A virtual private networking protocol that is part of IPv6 but is widely used now in IPv4.

IPv4 The current version of IP addressing based on 32-bit IP addresses.

IPv6 The next generation of IP addressing based on 64-bit IP addresses and having a number of enhancements over IPv4, such as automatic IP address configuration and better security.

ISDN Integrated Services Digital Network. An early member of DSL technology that can support up to 128-Kbps symmetrical service. It's routed through the ILEC's PSTN switches instead of through a DSLAM, as is the case for IDSL.

ISP Internet service provider. Any company that provides Internet access service.

ITC Independent Telephone Company. In the United States, a telephone company that was not owned by AT&T before the 1984 divestiture.

ITU International Telecommunications Union. The ITU is an international body of member countries that defines recommendations and standards relating to international telecommunications.

IXC Interexchange carrier. A long-distance telephone company.

Kbps Kilobits per second. A measurement of digital bandwidth where one Kbps equals one thousand bits per second. For example, 64 Kbps = 64,000 bps.

L2TP Layer Two Tunneling Protocol. An IETF protocol used for virtual private networking.

LAN Local Area Network. A data network that connects computers in an area usually within the confines of a building or floors within a building. A LAN enables users to share information and network resources, such as a printer or DSL CPE. Ethernet forms the basis of most local area networks.

last mile The local loop; the space between a local telephone company switching facility and the customer premises, a distance of about three miles.

latency A measure of the delay between the sending of a packet at the originating end of a connection and the reception of that packet at the destination end.

layer In the OSI network reference model, each layer performs a certain task to move the data from the sender to the receiver. Protocols within the layers define the tasks for the networks.

LEC Local Exchange Carrier. The telephone company that provides local telephone service. Since the Telecommunications Act of 1996, the term LEC has been replaced with ILEC.

loading coil A metallic, doughnut-shaped device used on local loops to extend their reach. Loading coils severely limit the bandwidth in digital communications.

local loop A generic term for the connection between the customer's premises and the telephone company's serving central office. The local loop is the pair of copper wires that connects the end user to the central office, which is the gateway to the telecommunications network.

MAC address Media Access Control address. The 48-bit defined number built into any Ethernet device connected to a LAN. This unique hardware address is represented as six octets, separated by colons, such as C0:3C:4E:00:10:8F. Bridges work at the MAC address level.

Mbps Million bits per second. A measurement of digital bandwidth where one Mbps equals one million bits per second.

MDF Main Distribution Frame. The point where all local loops are terminated at a CO.

MPOE Minimum Point of Entry. The place where phone lines first enter a customer's facility. The MPOE can be a network interface device or an inside wiring closet.

Mutliplexer Any one of a number of common devices used to combine and later split multiple telecommunications circuits into channels. DSL lines coming into the CO are mutliplexed to be carried over trunk lines.

MVL Multiple Virtual Lines. A DSL technology developed by Paradyne. MVL transforms a single copper loop into multiple virtual lines to support multiple independent services over the same line simultaneously.

NAT Network Address Translation. An Internet standard that allows your local network to use private IP addresses, which are not recognized on the Internet. The IP address used for the router is the only routable IP address. The computers behind the NAT can access the Internet through the router, but Internet users can't access the computers behind the router.

NDIS Network Driver Interface Specification. Developed by Microsoft to provide a common set of rules for network adapters to interface with operating systems.

NIC Network interface card. The hardware that forms the interface between the computer (or other network device) and not only the data communications network for the LAN but also the IP connection through a DSL bridge or router.

NID Network interface device. A device that terminates a copper pair from the serving central office at the user's destination. The NID is typically a small box installed on the exterior premises of the destination.

NNTP Network News Transport Protocol. The protocol that governs the transmission of network news, a threaded messaging system for posting messages to form newsgroup discussions.

NSP Network service provider. Any company that provides network services to subscribers.

OSI Open Systems Interconnection. An internationally accepted model of data communication protocols developed by OSI and ITU. The OSI Reference Model has seven layers of protocols used for networking.

packet A variable-sized unit of information that can be sent across a packet-switching network. A packet typically contains addressing information, error checking, user information, in addition to application data.

Packet CLEC A Competitive Local Exchange Carrier that focuses on providing data communication services instead of voice services. A term coined by Covad Communications to describe their data centric services.

packet filter The capability to search a packet to determine its destination and then route or block it accordingly. Routers perform this function to route TCP/IP data traffic.

packet-switched network A network that does not establish a dedicated path through the network for the duration of a session but instead transmits data in units called packets in a connectionless manner. Data streams are broken into packets at the front end of a transmission.

packet switching A data transmission method in which data is transferred by packets, or blocks of data. Packets are sent using a store-and-forward method across nodes in a network.

PC Card Term used for PCMCIA (Personal Computer Memory Card International Association) cards, which are the credit-card-size adapter cards you use in notebooks. A DSL modem for a notebook can be a PC Card.

PCI Peripheral Component Interconnect. A specification introduced by Intel that defines a local bus system that allows up to 10 PCI-compliant expansion cards in a PC. A de facto bus standard for today's PCs that has replaced the ISA (Industry Standard Architecture) bus.

peer-to-peer networks A network structure that creates an environment in which all computers are created equal: Each computer on the LAN can act as a server, a client, or both. No dedicated computer controls operations; everyone is expected to be a good networker and share resources (such as programs or data files) with others. If you use Windows 95, 98, or NT Workstation 4.0, you have built-in peer-to-peer networking capabilities.

PnP Plug-and-Play. A system for simplifying installation of hardware devices on a Microsoft Windows computer. Automates hardware recognition, driver installation, and system management.

POP3 Post Office Protocol, Version 3. The latest version of the Post Office Protocol, POP3 provides basic client/server features for handling e-mail. POP3 is supported by most e-mail client programs.

POTS Plain Old Telephone Service. A historical term for basic telephone voice service over two-wire copper loop and out to the PSTN.

PPP Point-to-Point Protocol. A communications protocol that allows a computer using TCP/IP to connect directly to the Internet through a dial-up connection. In Microsoft Windows, this type of connection is set up and controlled using Dial-Up Networking (DUN).

PPPoATM Point-to-Point Protocol over Asynchronous Transfer Mode. ATM is a high-speed switching technique used to transmit high volumes of voice, data, and video traffic. ATM operates at speeds ranging from 25 Mbps to 622 Mbps and is used mainly in telephone company backbone networks, although large organizations are also using ATM. Using PPP over ATM enables TCP/IP traffic to be carried over an ATM network all the way down to a computer without being translated. This configuration requires an ATM adapter card in each computer that connects to an ATM ADSL bridge or router.

PPPoE Point-to-Point-Protocol over Ethernet. An emerging standard that enables dial-up networking capabilities over Ethernet. PPPoE is a software driver that works with a NIC to create a dial-up session through the NIC and the LAN out through the DSL bridge or router. The significance of PPP over Ethernet has to do with making DSL service installation easier to set up for users and ISPs, as well as enabling users to access multiple network services from the same DSL connection.

protocol A set of rules that defines how different systems interoperate.

PSTN Public Switched Telephone Network. The network that provides global telephone service.

PUC Public Utility Commission. A United States government agency, usually at the state level, that regulates telecommunication companies and other utilities. PUCs manage the pricing of telecommunication services called tariffs. Telecommunications companies work with a PUC to bring DSL service to your area.

PVC Permanent Virtual Circuit. A static connection that has a predefined route. All information transferred across this connection traverses the same path throughout the network.

QoS Quality of Service. A premium class of data communications service in which the provider guarantees a level of service.

RADSL Rate-Adaptive Digital Subscriber Line. ADSL variant that offers automatic rate adaptation. Most ADSL is really RADSL, which allows the actual data transmission rates to adjust to line conditions and distance. Downstream speeds can reach up to 8 Mbps, and upstream speed can reach up to around 1 Mbps. RADSL supports both asymmetrical and symmetrical data transmission.

RBOCs Regional Bell Operating Companies. The seven original regional Bell operating companies that provided local telephone service and were formed as a result of the AT&T divestiture. They are Ameritech, Bell Atlantic, Bell South, NYNEX, Pacific Bell, Southwestern Bell, and US WEST.

RJ-11 A standard modular connector (jack or plug) that supports two pairs of wires (four wires). Commonly used for most PSTN CPE (telephones, fax machines, and modems).

RJ-45 A standard modular connector that can support up to four pairs of wires (eight wires). RJ-45 connectors are used with Category 5 cabling used with 10Base-T or 100Base-T cabling. This cabling is also used in business locations with more sophisticated voice communication systems.

router A device that routes data between networks through IP addressing information contained in the header of the IP packet. A router forwards packets to other routers until the packets reach their destination. Routers form the basis of IP networking.

SDSL Symmetrical Digital Subscriber Line. A member of the DSL family that is being widely deployed by CLECs. SDSL supports symmetrical service at 160 Kbps to 2.3 Mbps but does not support POTS connections. SDSL uses the stable 2B1Q line coding scheme, which makes it a quiet neighbor in cable bundles. SDSL can reach up to 23,000 feet from the CO.

server A host that makes an application or a service available to other hosts, typically clients. For example, a Web server relays information to Web browser clients.

signal processing When an electrical signal is transmitted over some medium for communications usage.

SMTP Simple Mail Transfer Protocol. SMTP is the protocol for Internet e-mail that transfers e-mail messages among computers. SMTP uses a store-and-forward system to move e-mail messages to their final destination.

SNMP Simple Network Management Protocol. A protocol used for remote management of internetworking devices and networks. SNMP is used by a management station to monitor and configure network devices such as routers, bridges, hubs, and hosts.

splitter A device used to separate POTS service from the ADSL data service at a customer's premises. The CO side of the DSL connection also has a POTS splitter.

SSL Secure Sockets Layer. SSL version 2 provides security by allowing applications to encrypt data that goes from a client, such as Web browser, to a matching server. (Encrypting your data means converting it to a secret code.) SSL version 3 allows the server to authenticate that the client is who it says it is.

standard A set of technical specifications used to establish uniformity in software, hardware, and data communications.

static IP addressing An assigned IP address used to connect to a TCP/IP network. The IP address stays with the specific host or network device. Typically used with routable public IP addresses so that a particular host can be reached by its assigned static IP address and any domain name associated with the IP address.

STP Shielded twisted pair. A shielded form of the twisted-pair wiring used for 10Base-T and 100Base-T LANs. STP has a foil or wire braid wrapped around the individual wires to provide better protection against electromagnetic interference. STP uses different connectors than UTP, is more expensive than UTP, and requires careful grounding to work properly.

subnet A portion of a network. Each subnet within a network shares a common network address and is uniquely identified by a subnetwork number.

subnet mask A 32-bit number used to separate the network and host sections of an IP address. A subnet mask subdivides an IP network into smaller pieces. An example of a subnet mask address might be 255.255.255.248 for an 8 IP address network.

T-1 A North American standard for communicating at 1.54 Mbps. A T-1 line has the capacity for 24 voice and data channels at 64 Kbps each. The benchmark in leased, digital line service from ILECs.

T-3 A North American standard for communicating at speeds of 44 Mbps. A T-3 line has 672 channels for voice and data at 64 Kbps each.

TCP Transmission Control Protocol. One of two principal components of the TCP/IP protocol suite. TCP puts data into packets and provides packet delivery across the network, ensuring that packets are not lost in transmission and arrive in order.

TCP/IP Transmission Control Protocol/Internet Protocol. TCP/IP is the suite of protocols that define the basis of the Internet. It provides communications across interconnected networks between computers with diverse hardware and operating systems.

TCP/IP stack The software that allows a computer to communicate through TCP/IP. Stack refers to the fact that there are five layers of protocols operating on a TCP/IP network.

Telecommunications Act of 1996 A piece of legislation passed by the U.S. Congress that is trying to help open up the local telecommunications and cable industries to competition. The goal is to give consumers more choices at lower costs. Although by no means a perfect piece of legislation, it's already bearing fruit in the deployment of DSL.

telephony The science of transmitting voice, data, and video over a distance greater than you can transmit by shouting.

telnet A terminal-emulation protocol that allows you to access computers and network devices through TCP/IP.

trunk A communications link between two switching systems, such as PSTN switches.

twisted pair A cable comprised of pairs of wires twisted around each other to help cancel out interference. This is the common form of copper cabling used for telephony and data communications.

UART Universal Asynchronous Receiver/Transceiver. The older serial port architecture for data communications that is limited to 115-Kbps capacity. UART is being replaced with USB (Universal Serial Bus).

UAWG Universal ADSL Working Group. A consortium of telecommunications service providers, PC and network equipment vendors, and DSL equipment vendors supporting the emergence of the G.Lite DSL standard that forms the basis of UADSL.

upstream The direction of information flow on a data communications link. The faster the upstream speed, the faster data can move from your local network or PC to the Internet. If you run a Web server or any type of TCP/IP server, it will effect your upstream capacity.

USB Universal Serial Bus. A new data communications port installed on most newer PCs to replace the UART serial port. USB ports are easier to use for plugging in peripherals and support data communication speeds up to 12 Mbps.

UTP Unshielded twisted pair. Cabling used for 10Base-T and 100Base-T LANs. UTP consists of pairs of copper wires twisted around each other and covered by plastic insulation. UTP is by far the most popular cabling used for LANs.

VDSL Very-high-bit-rate Digital Subscriber Line. An ultra-high-speed DSL flavor that can deliver data communications at speeds up to 52 Mbps in close proximity to a CO.

VPN Virtual private network. A way that private data can safely pass over a public network, such as the Internet. The data traveling between the two hosts are encrypted for privacy using both hardware and software solutions.

WAN Wide area network. A data network typically extending a LAN outside a building, over a data communications link to another network in another location. A WAN typically uses common carrier lines. The jump between a LAN and a WAN is made through a bridge or a router.

Web hosting The placement of a Web server off-site, usually with an ISP or a Web hosting service company that operates all the Web server infrastructure for you.

WinSock A program that conforms to a set of specifications called the Windows Socket API (Application Programming Interface). WinSock controls the link between Microsoft Windows and a TCP/IP program.

xDSL A generic term used to refer to the entire family of DSL technologies. The x is a placeholder for A in ADSL, S in SDSL, and so on. These DSL technologies are ADSL, HDSL, IDSL, SDSL, RADSL, VDSL, and UADSL.

Index

Notes

Notes

Notes

Think You Can t Afford High Speed Access?

Get ConcentricDSL

Don t let speed be a barrier to your business.

With ConcentricDSL™ you get Internet and LAN access that s fast, easy, and affordable. It s up to **50** times faster than an analog modem and **10** times faster than ISDN. So if you re tired of waiting for data transfers, get ConcentricDSL.

concentric network

1-888-493-6232
www.concentric.net/dsl